Awesome 'Possum

Volume 3

Edited by Angela Boyle

Awesome 'Possum, Volume 3
Published by Flying Dodo Publications

Flying Dodo Publications
ISBN 978-0-9970111-2-8
10 9 8 7 6 5 4 3 2 1   16 17 18 19

Editior Angela Boyle
Additional scientific editor: Abe Olson
Design and layout: Angela Boyle
Heading font: Birmingham by Paul Lloyd
Animals sillouhette font: La monde de Victor by Philippe Andrevon

Flying Dodo Publications
White River Junction, VT
angela@flyingdodopublications.com
flyingdodopublications.com

First printing, December 2016

Printed in the United States of America.

Table of Contents

Introduction

by Ursula Vernon

So I have a hobby that some people may consider peculiar. I like to take photographs of moths.

These are not good photographs, I hasten to add. This is not high art. I am not trying to get moth glamour shots here. I just want to get the field marks. Then I email the photo off to a website where they identify it and drop a little pin in a map to indicate the sighting.

I'm not an entomologist. I came to bugs late in life. This is citizen science in its most basic form, with somebody who didn't know a leafhopper from a luna moth when she started out.

Over the course of a couple years, I've ID'd nearly three hundred moth species in my garden. The first few were on a whim, but then I got hooked. I think it was the names that did it. How can you not love "The False Underwing" and the "Intractable Quaker" and the "Stinging Rose"? Some seem like bizarre projection—what makes the "Grateful Midget" grateful, anyway? And some are almost sinister, like "The Laugher" or the "Hickory Horned Devil."

I fell in love, and the love persisted through a luna moth flying into my face like a terrified bird and the discovery of asp caterpillars lurking in the garden, which look like tiny toupees and sting like death.

The first time I found a moth and logged the first report of that species in my whole state, I had to go lie down for a few minutes with a cold cloth over my eyes. It didn't seem possible that I had found something glommed onto my doorframe and nobody else had reported one.

But it's not that they were rare. It's only that hardly anybody was looking.

It's easy, in our modern world, to perceive nature as something happening out there. Nature is lions on the savannah, penguins trudging through Antarctica, grizzly bears flipping salmon in Alaskan streams. All glorious and magnificent and impressive, of course, but unless you happen to actually live in one of those areas, there's an unreality to those images. They're off in the distance, probably being narrated by Sir David Attenborough.

(This is not to disparage Sir David Attenborough. In fact, nobody better disparage him in my hearing. I will fight you.)

But when we put nature out there, happening somewhere else, we lose sight of the fact that nature is all around us and we are interacting with it all the time.

Sure, lions are awesome. I love a good penguin documentary as much as the next person. But at the end of the day, most people's experience with nature will be the nature that lives in our own backyards: the urban hawk and the suburban fox, the pigeons perched on power lines and the squirrels using telephone wires as their own personal highways. Even the moths at the porch light.

Depending on where you live, you may have hedgehogs in the leaf piles or armadillos stomping through the brush, but it's almost a given that there will be something.

It's all too easy to overlook these locals, or to notice only a charismatic few. We take squirrels for granted, never stopping to notice the incredible engineering in their joints. We grumble over opossums in the trash can, unaware that these weird, primitive marsupials have held for millions of years on a continent dominated by placental mammals. We marvel at the flight of the hummingbird and miss the hummingbird hawkmoth; we point at the bald eagle in delight without realizing that it's probably stealing its meals from the hardworking local osprey or scavenging dead fish cast up on shore.

When we look a little more closely at the nature that is all around us, even in cities and suburbs, we find a weird and delightful world, of which humans form an inextricable part. Cicadas hatch under our feet and emerge in broods to fill the air with buzzing. The local food chain expands to include trash cans and Chihuahuas (and one time, my stepmom's miniature poodle). Eastern coyotes interbreed with dogs. Cane toads cover golf courses, and ants form supercolonies in the California suburbs.

The volume you hold in your hands brings all that home. Some of the comics are subversive and funny—I'm still laughing over the rat in swimwear—and some are sweet and touched with wonder, like the history of fossil discovery in Dudley.

It's natural science for the modern age: hilarious, weird, thought-provoking, and right in our own backyards.

The Handy Opossum
FRONT FOOT
by Angela Boyle
No nail
HIND FOOT
Opossums have opossable thumbs on their hind feet.
Primates are the only mammals other than opossums with an opossable thumb on their feet.
Opossums also use their prehensile tails for grabbing, yet another adaptation that aids in climbing. They can hang from their tails for only a shortwhile.
They can even carry bunches of grass and leaves in their tails to use in making their dens.
Their sharp claws dig into bark for a better grip.
They might be okay climbers, but they are slow amblers with a top land speed of 4–7 mph. (6.4– 11.2 Kph).
(Humans walk at 3mph)

IRIS YAN

Apis mellifera & Varroa destructor:
Detection & Treatment of Varroa Mites in the Honeybee Hive
Text & Graphics by
William Scavone, MA, CMI
Oh no!
What's wrong, daddy?
I think it may be time to treat our honeybees for Varroa mites.
What's a Varroa mite?
The Varroa mite, Varroa destructor, came to the United States in 1987 and had, by 1998, wiped out almost all of the feral bee colonies.
It is currently the most severe pest of managed honeybees worldwide, especially the European honeybee, Apis mellifera.
Scavone
The Varroa mite is a honeybee parasite that latches onto the bees and sucks their blood—or hemolymph.
Eww.
Gross!
Like a tick?
Yep, except that it only feeds on bees.
See this little speck on the abdomen of this worker bee?
That's a Varroa mite.
It's a tiny little bug with eight legs, about the size of a pinhead, and extremely hard to see on a bee.
Just like a tick, the female sucks the blood of her host in order to raise offspring.

But while a tick that hops on us out in the field is tiny compared to us...
DEER TICK
NYMPH
ADULT
(ACTUAL SIZE)
...the Varroa mite is huge compared to a honeybee.
4x
Imagine a tick the size of a cat hanging onto your back and sucking your blood!.
Eww...
!
Yikes!
Get it off!
And just as ticks transmit diseases to people – like Lyme disease and Rocky Mountain spotted fever – Varroa mites carry viruses and other pathogens that can make the bees sick.
Look at this young bee I just found. See how her wings are small and shriveled?
She has deformed wing virus (D W V), which is transmitted by the mites.
She won't be able to fly and forage for food, and will probably die within 48 hours.
Bees with mites also have suppressed immune systems and are susceptible to other diseases that can lead to colony collapse.*
*Colony collapse disorder was first identified by beekeepers in 2012.
For seemingly unexplained causes, they would find otherwise healthy hives empty of bees.
Signs like D W V mean we have to treat right away and check our other hives for mites.
Poor little honeybee.
That's why we have to nip this in the bud!

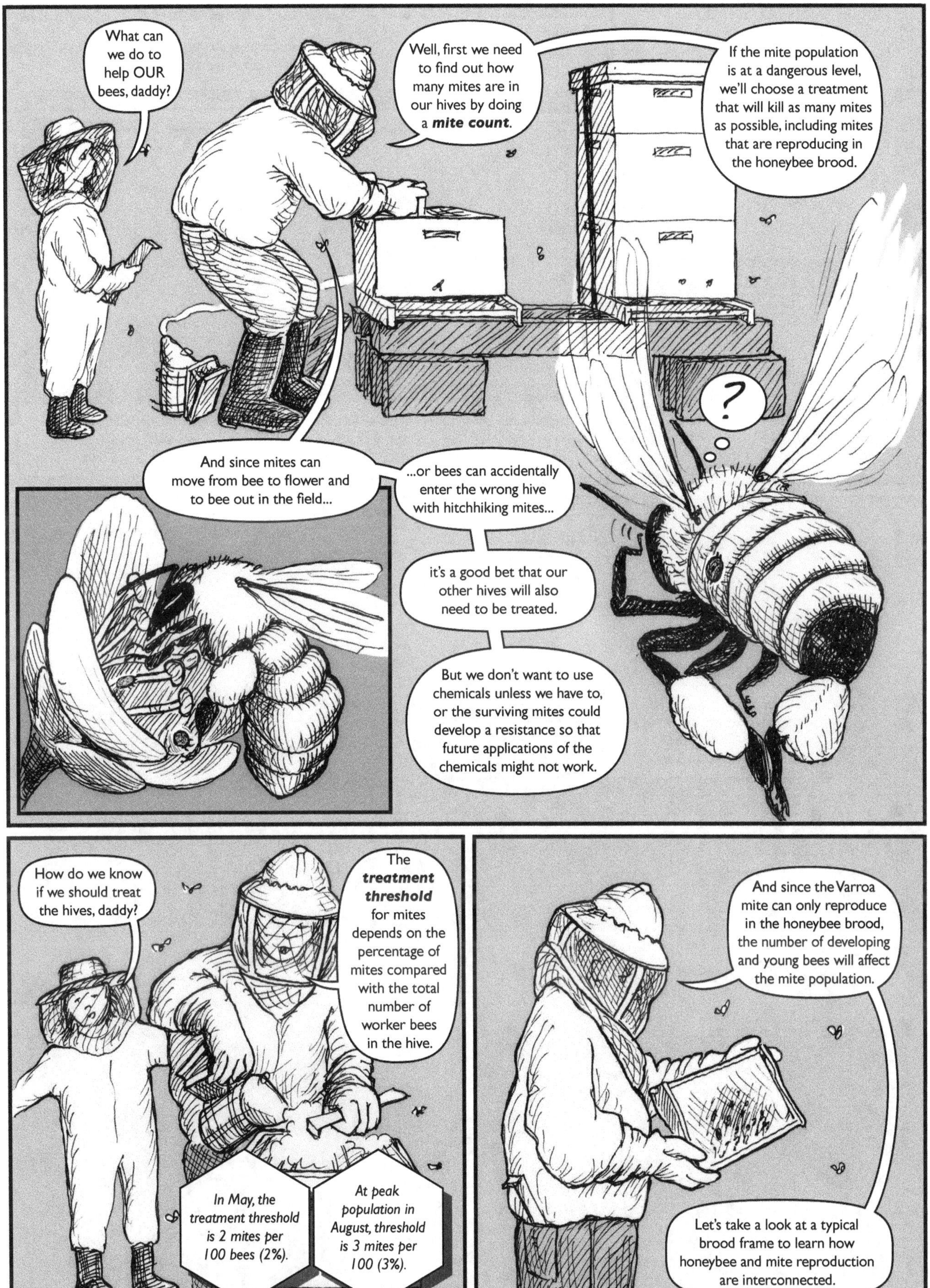
What can we do to help OUR bees, daddy?
Well, first we need to find out how many mites are in our hives by doing a **mite count**.
If the mite population is at a dangerous level, we'll choose a treatment that will kill as many mites as possible, including mites that are reproducing in the honeybee brood.
?
And since mites can move from bee to flower and to bee out in the field...
...or bees can accidentally enter the wrong hive with hitchhiking mites...
it's a good bet that our other hives will also need to be treated.
But we don't want to use chemicals unless we have to, or the surviving mites could develop a resistance so that future applications of the chemicals might not work.
How do we know if we should treat the hives, daddy?
The ***treatment threshold*** for mites depends on the percentage of mites compared with the total number of worker bees in the hive.
In May, the treatment threshold is 2 mites per 100 bees (2%).
At peak population in August, threshold is 3 mites per 100 (3%).
And since the Varroa mite can only reproduce in the honeybee brood, the number of developing and young bees will affect the mite population.
Let's take a look at a typical brood frame to learn how honeybee and mite reproduction are interconnected.

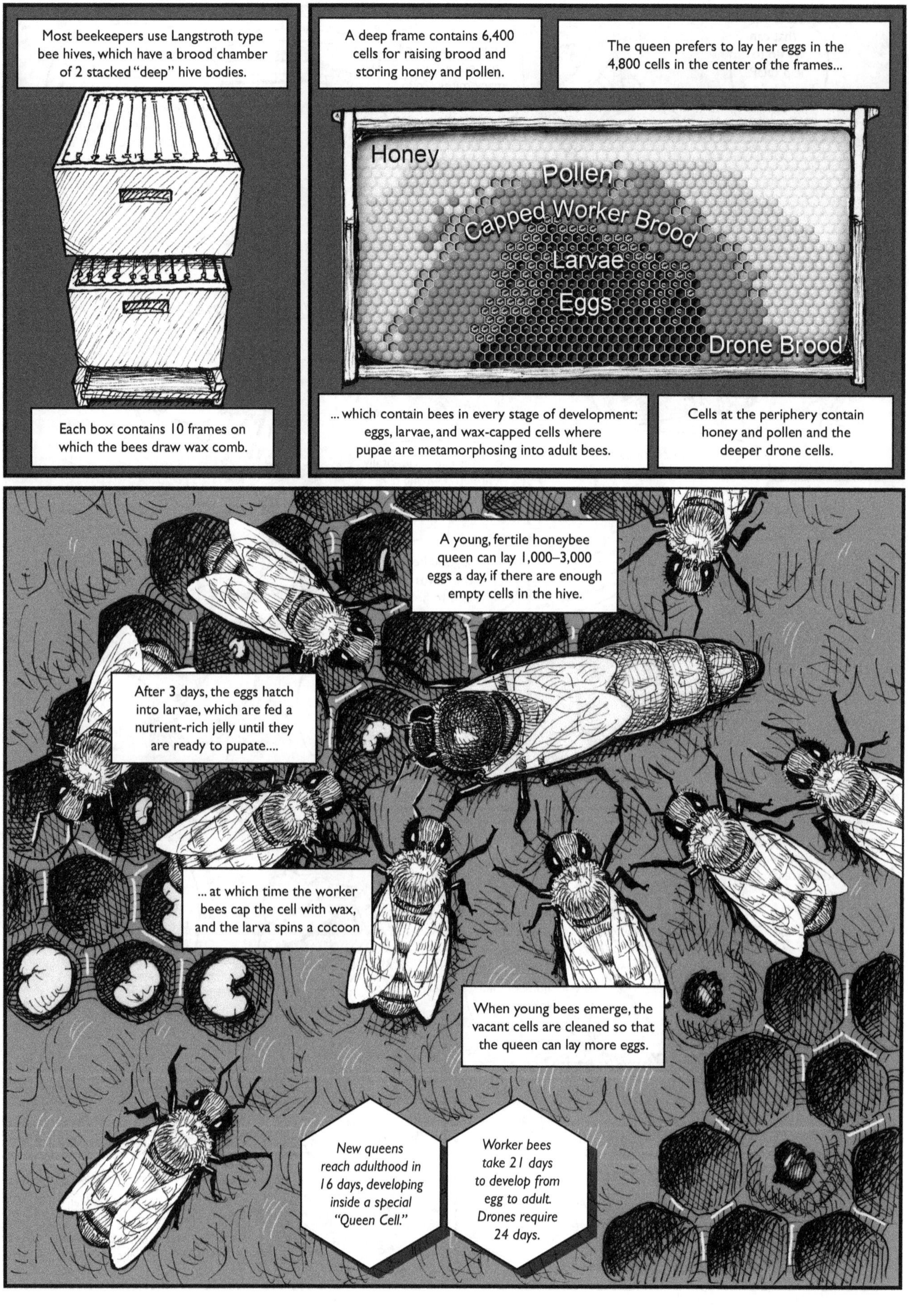
Most beekeepers use Langstroth type bee hives, which have a brood chamber of 2 stacked "deep" hive bodies.
Each box contains 10 frames on which the bees draw wax comb.
A deep frame contains 6,400 cells for raising brood and storing honey and pollen.
The queen prefers to lay her eggs in the 4,800 cells in the center of the frames...
Honey
Pollen
Capped Worker Brood
Larvae
Eggs
Drone Brood
... which contain bees in every stage of development: eggs, larvae, and wax-capped cells where pupae are metamorphosing into adult bees.
Cells at the periphery contain honey and pollen and the deeper drone cells.
A young, fertile honeybee queen can lay 1,000–3,000 eggs a day, if there are enough empty cells in the hive.
After 3 days, the eggs hatch into larvae, which are fed a nutrient-rich jelly until they are ready to pupate....
... at which time the worker bees cap the cell with wax, and the larva spins a cocoon
When young bees emerge, the vacant cells are cleaned so that the queen can lay more eggs.
New queens reach adulthood in 16 days, developing inside a special "Queen Cell."
Worker bees take 21 days to develop from egg to adult. Drones require 24 days.

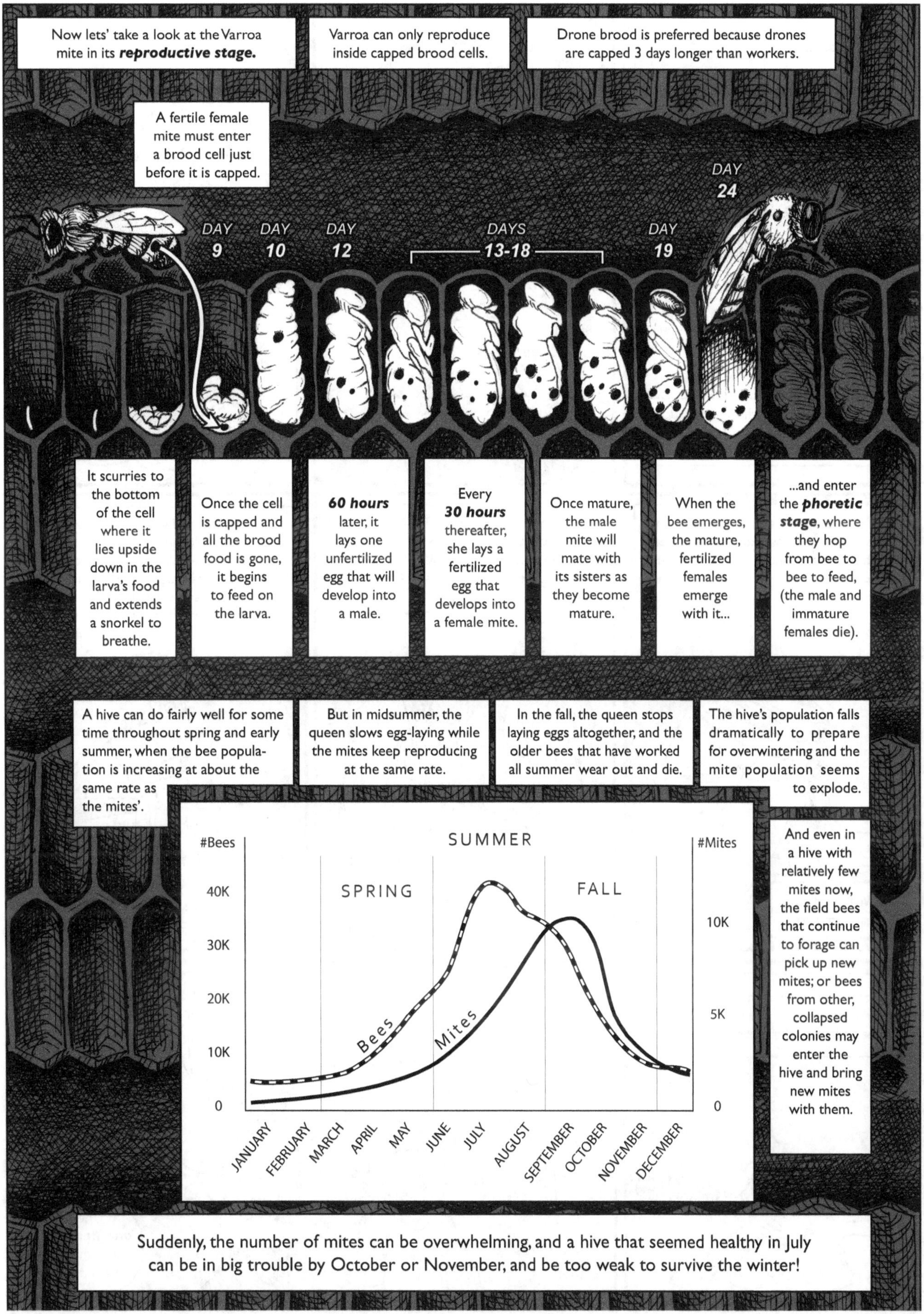
Now lets' take a look at the Varroa mite in its ***reproductive stage.***
Varroa can only reproduce inside capped brood cells.
Drone brood is preferred because drones are capped 3 days longer than workers.
A fertile female mite must enter a brood cell just before it is capped.
DAY 9
DAY 10
DAY 12
DAYS 13-18
DAY 19
DAY 24
It scurries to the bottom of the cell where it lies upside down in the larva's food and extends a snorkel to breathe.
Once the cell is capped and all the brood food is gone, it begins to feed on the larva.
60 hours later, it lays one unfertilized egg that will develop into a male.
Every ***30 hours*** thereafter, she lays a fertilized egg that develops into a female mite.
Once mature, the male mite will mate with its sisters as they become mature.
When the bee emerges, the mature, fertilized females emerge with it...
...and enter the ***phoretic stage***, where they hop from bee to bee to feed, (the male and immature females die).
A hive can do fairly well for some time throughout spring and early summer, when the bee population is increasing at about the same rate as the mites'.
But in midsummer, the queen slows egg-laying while the mites keep reproducing at the same rate.
In the fall, the queen stops laying eggs altogether, and the older bees that have worked all summer wear out and die.
The hive's population falls dramatically to prepare for overwintering and the mite population seems to explode.
And even in a hive with relatively few mites now, the field bees that continue to forage can pick up new mites; or bees from other, collapsed colonies may enter the hive and bring new mites with them.
#Bees
#Mites
SUMMER
SPRING
FALL
40K
30K
20K
10K
0
10K
5K
0
Bees
Mites
JANUARY
FEBRUARY
MARCH
APRIL
MAY
JUNE
JULY
AUGUST
SEPTEMBER
OCTOBER
NOVEMBER
DECEMBER
Suddenly, the number of mites can be overwhelming, and a hive that seemed healthy in July can be in big trouble by October or November, and be too weak to survive the winter!

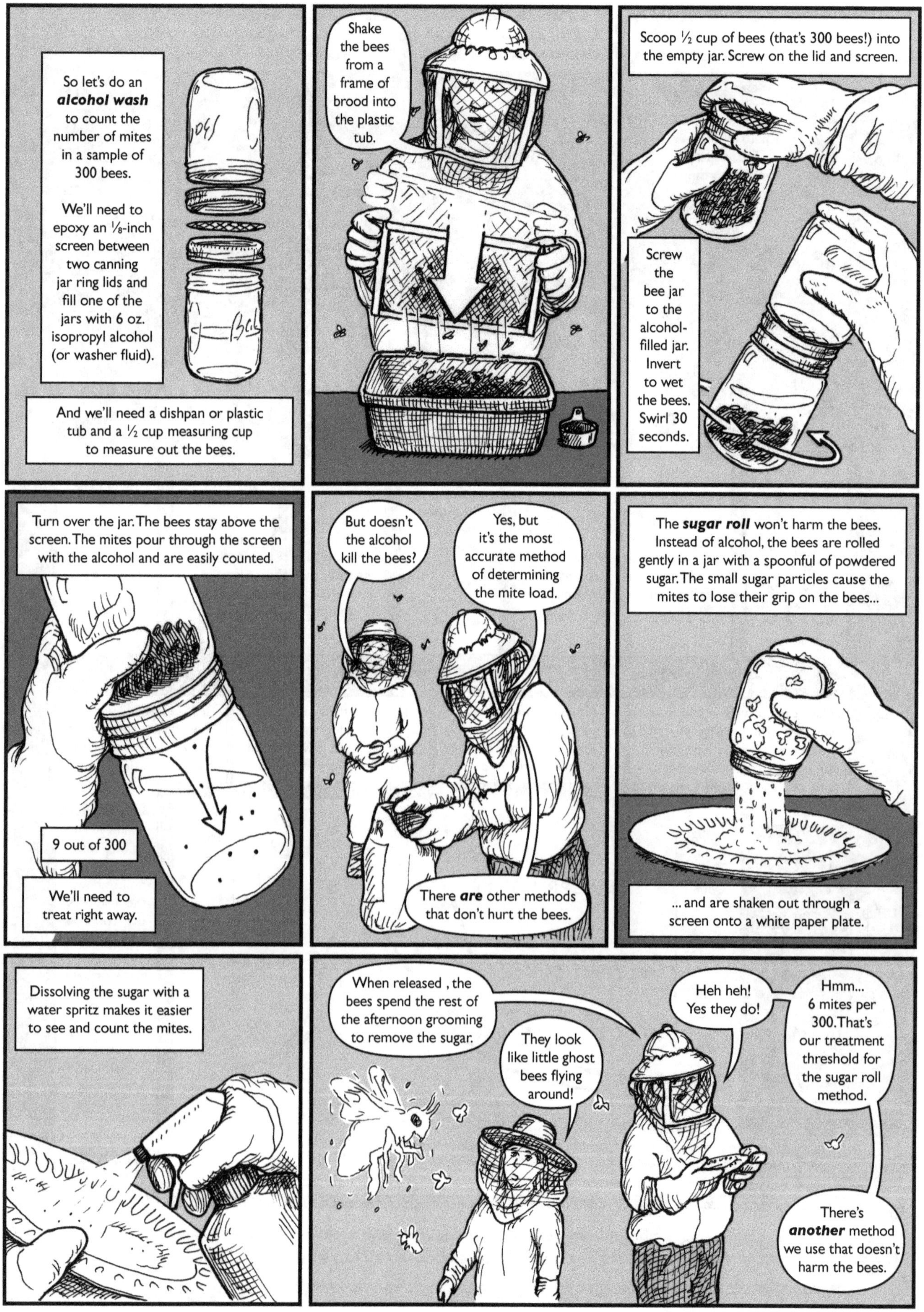
So let's do an ***alcohol wash*** to count the number of mites in a sample of 300 bees.
We'll need to epoxy an ⅛-inch screen between two canning jar ring lids and fill one of the jars with 6 oz. isopropyl alcohol (or washer fluid).
And we'll need a dishpan or plastic tub and a ½ cup measuring cup to measure out the bees.
Shake the bees from a frame of brood into the plastic tub.
Scoop ½ cup of bees (that's 300 bees!) into the empty jar. Screw on the lid and screen.
Screw the bee jar to the alcohol-filled jar. Invert to wet the bees. Swirl 30 seconds.
Turn over the jar. The bees stay above the screen. The mites pour through the screen with the alcohol and are easily counted.
9 out of 300
We'll need to treat right away.
But doesn't the alcohol kill the bees?
Yes, but it's the most accurate method of determining the mite load.
There ***are*** other methods that don't hurt the bees.
The ***sugar roll*** won't harm the bees. Instead of alcohol, the bees are rolled gently in a jar with a spoonful of powdered sugar. The small sugar particles cause the mites to lose their grip on the bees...
... and are shaken out through a screen onto a white paper plate.
Dissolving the sugar with a water spritz makes it easier to see and count the mites.
When released , the bees spend the rest of the afternoon grooming to remove the sugar.
They look like little ghost bees flying around!
Heh heh! Yes they do!
Hmm... 6 mites per 300. That's our treatment threshold for the sugar roll method.
There's ***another*** method we use that doesn't harm the bees.

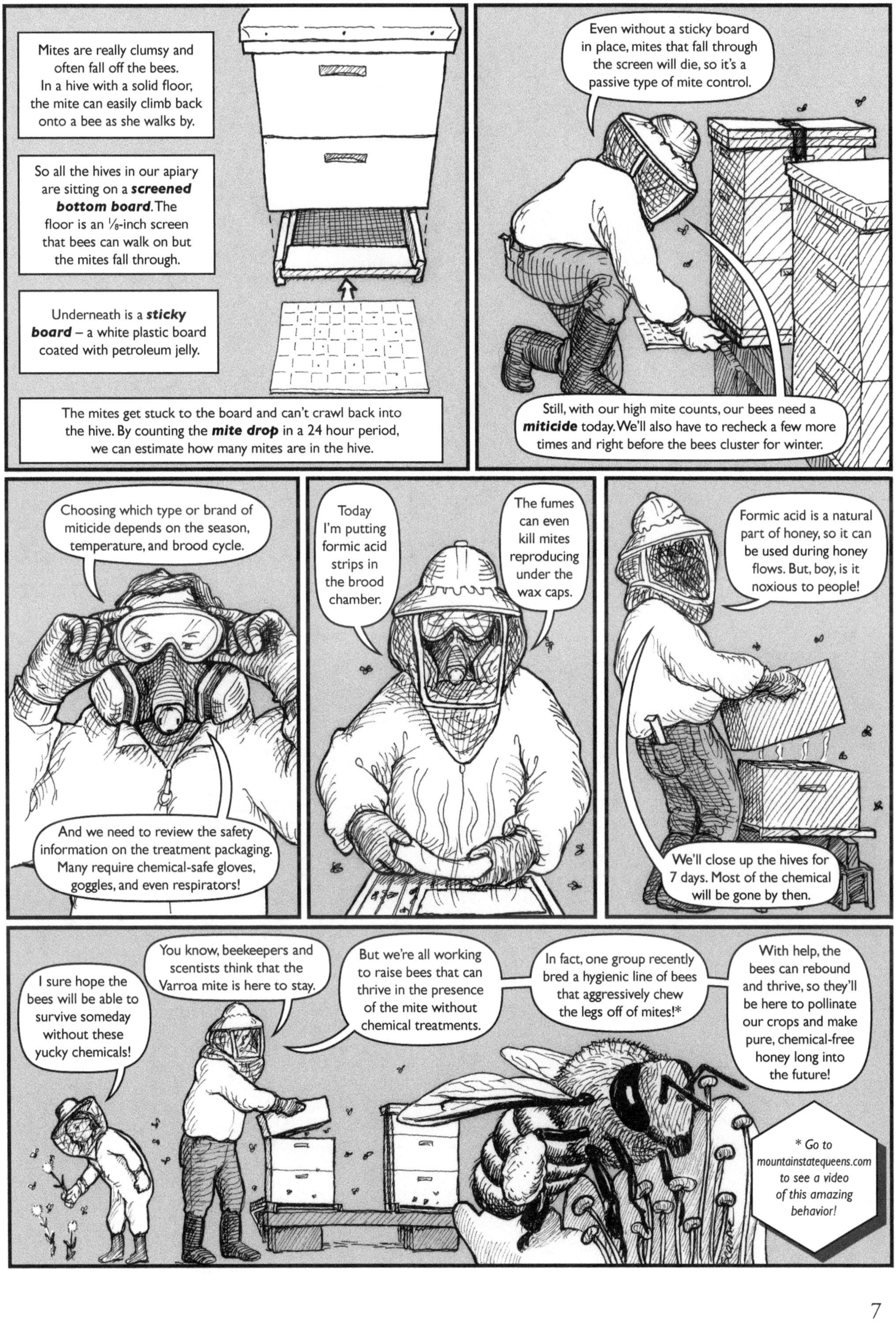
Mites are really clumsy and often fall off the bees. In a hive with a solid floor, the mite can easily climb back onto a bee as she walks by.
So all the hives in our apiary are sitting on a screened bottom board. The floor is an ⅛-inch screen that bees can walk on but the mites fall through.
Underneath is a sticky board – a white plastic board coated with petroleum jelly.
The mites get stuck to the board and can't crawl back into the hive. By counting the mite drop in a 24 hour period, we can estimate how many mites are in the hive.
Even without a sticky board in place, mites that fall through the screen will die, so it's a passive type of mite control.
Still, with our high mite counts, our bees need a miticide today. We'll also have to recheck a few more times and right before the bees cluster for winter.
Choosing which type or brand of miticide depends on the season, temperature, and brood cycle.
And we need to review the safety information on the treatment packaging. Many require chemical-safe gloves, goggles, and even respirators!
Today I'm putting formic acid strips in the brood chamber.
The fumes can even kill mites reproducing under the wax caps.
Formic acid is a natural part of honey, so it can be used during honey flows. But, boy, is it noxious to people!
We'll close up the hives for 7 days. Most of the chemical will be gone by then.
I sure hope the bees will be able to survive someday without these yucky chemicals!
You know, beekeepers and scientists think that the Varroa mite is here to stay.
But we're all working to raise bees that can thrive in the presence of the mite without chemical treatments.
In fact, one group recently bred a hygienic line of bees that aggressively chew the legs off of mites!*
With help, the bees can rebound and thrive, so they'll be here to pollinate our crops and make pure, chemical-free honey long into the future!
* Go to mountainstatequeens.com to see a video of this amazing behavior!

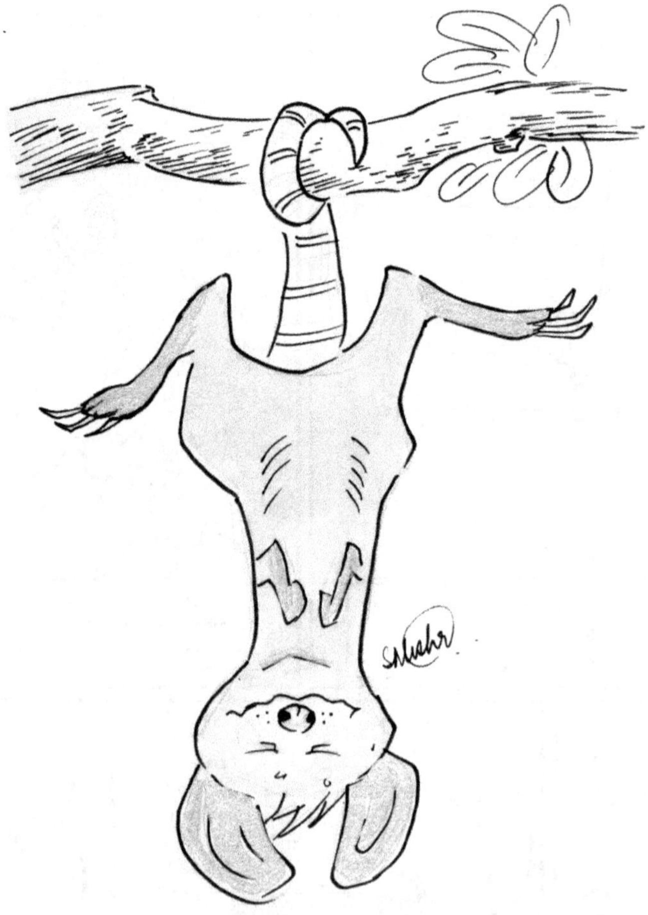

Shashwat Mishra

Don't Tread
On Us
by Spratty

RATTLESNAKES
ARE A FAMILY OF VIPERID SNAKES FOUND DISTRIBUTED THROUGHOUT NORTH & SOUTH AMERICA. OF THE 29 SPECIES OF RATTLESNAKE, 14 OF THEM ARE FOUND WITHIN THE CONTINENTAL UNITED STATES.
THEY ARE BEST KNOWN FOR THE FOLLOWING CHARACTERISTICS:
-THEIR VENOMOUS BITE
-SPECIALIZED FACIAL PIT ORGANS THAT DETECT INFRARED
-DISTINCTIVE WARNING SYSTEM ISSUED BY THEIR RATTLING TAILS
BZZZZ
ZZZZZZ
ZZZZ
-AND THEIR PARENTING SKILLS AND SOCIAL INTERACTIONS

RATTLESNAKES ARE OVOVIVIPAROUS (MEANING THE EGGS ARE INCUBATED INSIDE THE BODY OF THE MOTHER UNTIL THEY HATCH) GIVING BIRTH TO LITTERS OF UP TO 20 LIVE YOUNG IN THE SPRING AND SUMMER EVERY 2-3 YEARS.
NEWBORN RATTLESNAKES HAVE POOR EYESIGHT DUE TO THEIR SOON-TO-BE-SHED SKINS AND LACK THE STRIKING STRENGTH OF AN ADULT.
THEY ARE ALSO BORN WITHOUT THEIR SIGNATURE "RATTLE" POSESSING ONLY A STRUCTURE KNOWN AS A "BUTTON".
ALL 14 SPECIES OF RATTLESNAKE FOUND IN THE U.S. HAVE BEEN OBSERVED EXHIBITING A BEHAVIOR TERMED BY SCIENTISTS AS "MATERNAL ATTENDANCE." MOTHER RATTLESNAKES EXHIBIT MATERNAL ATTENDANCE BY BASKING WITH THEIR YOUNG. TO HELP KEEP THEM WARM WITH RADIATED BODY HEAT.
THEY PREVENT YOUNG FROM VENTURING INTO AREAS AWAY FROM THEIR DEN THAT MAY BE POTENTIALLY DANGEROUS.
THEY USE THEIR OWN RATTLES AND DEFENSIVE POSTURING TO WARD OFF PREDATORS SO THE BABIES CAN ESCAPE.
AFTER THE YOUNG SNAKES SHED THEIR SKINS FOR THE FIRST TIME (TYPICALLY 7-10 DAYS AFTER BIRTH) AND GAIN THE FIRST SCALES ON THEIR RATTLES, THEY VENTURE OFF ON THEIR OWN.

IN ORDER TO VERIFY THE BOND BETWEEN MOTHER RATTLESNAKES AND THEIR OFFSPRING, A TEAM OF RESEARCHERS LED BY PETER MAY OF STETSON UNIVERSITY IN DE LAND, FLORIDA DECIDED TO RUN A SERIES OF TESTS ON A GROUP OF 16 RATTLESNAKE FAMILIES.

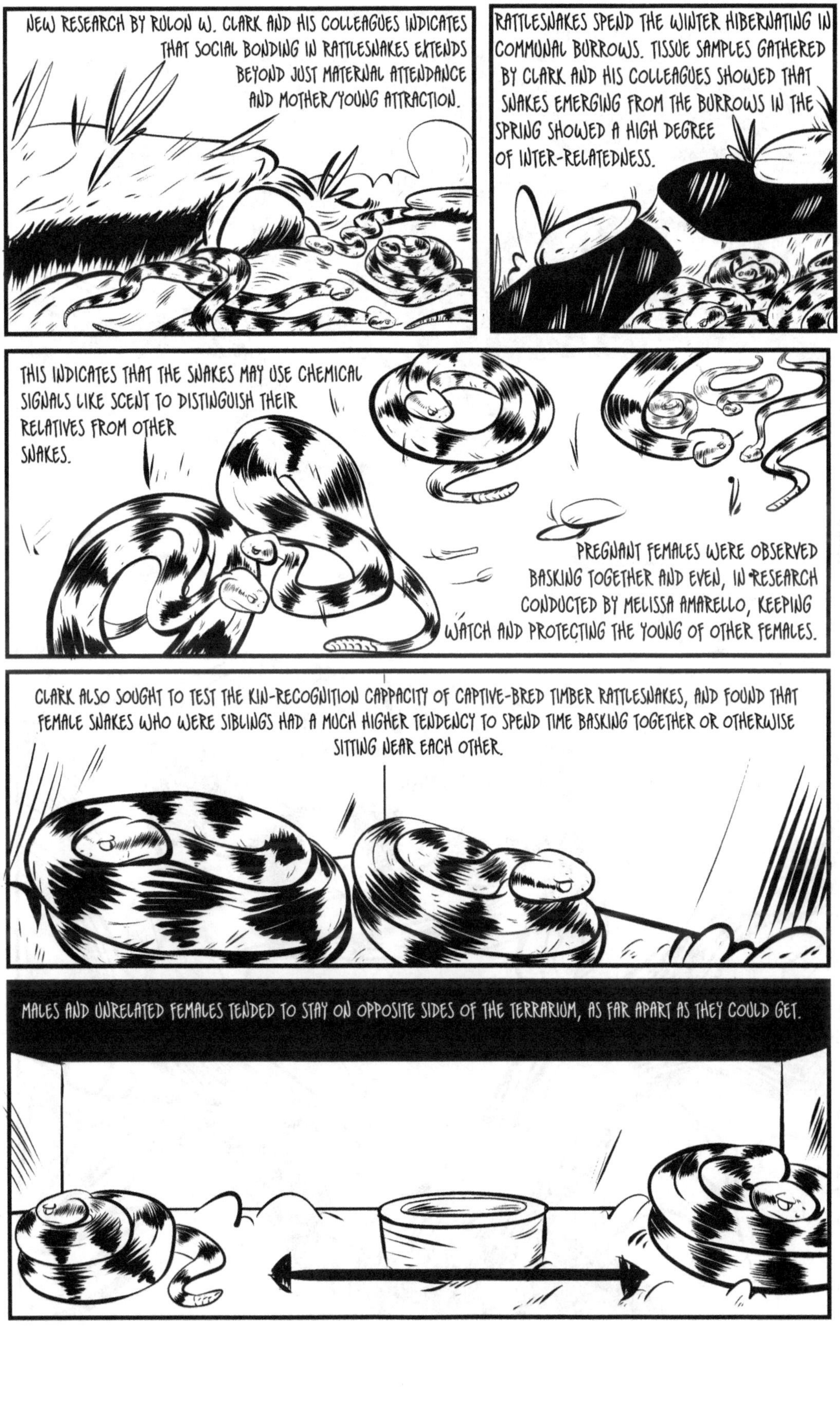
New research by Rulon W. Clark and his colleagues indicates that social bonding in rattlesnakes extends beyond just maternal attendance and mother/young attraction.
Rattlesnakes spend the winter hibernating in communal burrows. Tissue samples gathered by Clark and his colleagues showed that snakes emerging from the burrows in the spring showed a high degree of inter-relatedness.
This indicates that the snakes may use chemical signals like scent to distinguish their relatives from other snakes.
Pregnant females were observed basking together and even, in research conducted by Melissa Amarello, keeping watch and protecting the young of other females.
Clark also sought to test the kin-recognition cappacity of captive-bred timber rattlesnakes, and found that female snakes who were siblings had a much higher tendency to spend time basking together or otherwise sitting near each other.
Males and unrelated females tended to stay on opposite sides of the terrarium, as far apart as they could get.

LONG THOUGHT TO BE COMPLETELY ASOCIAL, RATTLESNAKES MAY, BASED ON THESE NEW FINDINGS, ACTUALLY BE FAMILY-ORIENTED ANIMALS WITH SUBTLE BUT COMPLEX SOCIAL LIVES.
THE DISCOVERY OF THEIR USE OF CHEMICAL OR SCENT-BASED CUES AND COMMUNICATION OPENS UP A NEW LOOK AT THE WAY ANIMALS SOCIALIZE AND MAY SERVE TO CHANGE THE WAY WE LOOK AT SNAKES AND THE SIGNIFICANCE OF REPTILE INTELLIGENCE.

Holy Fire

THE SECRET HISTORY OF ERGOT

MOSS BASTILLE

HEY THERE! IT'S ME, ANTHONY, PATRON SAINT OF HORRIBLE SKIN DISEASES!
NOW, I KNOW WHAT YOU'RE THINKING. "SKIN DISEASES? LIKE, RASHES?"
NOT QUITE, MY FRIEND! THERE'S A LONG AND GRUESOME HISTORY BEHIND MY PATRONAGE. LET ME TELL YOU ABOUT IT...

IN MEDIEVAL EUROPE, CHRISTIANITY WAS BOTH A PART OF EVERYDAY LIFE AND A SOURCE OF FEAR.
THE DEVIL WORKED CONSTANTLY TO BRING HUMANITY TO RUIN AND GOD WAS NOT MUCH KINDER.
HE COULD SAVE SOULS, BUT HE COULD ALSO DAMN THEM.
SO IT'S NO SURPIRSE THAT THE APPEARANCE OF A MYSTERIOUS PLAGUE WOULD BE BLAMED ON SUPERNATURAL OR DIVINE FORCES. THE "DANCING SICKNESS" CAUSED ITS VICTIMS TO CONTORT THEIR BODIES IN PAIN; THESE SPASMS WERE OFTEN ACCOMPANIED BY TERRIFYING VISIONS OF WILD ANIMALS AND FLAMES.
IT CAME TO BE SEEN AS DIVINE RETRIBUTION, EARNING THE NAME "HOLY FIRE" FOR THE BURNING SENSATIONS IT CAUSED.
IN 1039, AN OUTBREAK IN FRANCE INSPIRED GASTON DE LA VALLOIRE TO OPEN A HOSPITAL FOR VICTIMS OF THE DISEASE. HE DEDICATED IT TO ST. ANTHONY. IN THIS WAY, "HOLY FIRE" ALSO BECAME KNOWN AS "ST. ANTHONY'S FIRE."
IN TIME, THE MONKS OF THE ORDER OF ST. ANTHONY WOULD BUILD MORE THAN 370 SUCH HOSPITALS.

IN 1670, A FRENCH PHYSICIAN, DR. THUILLER, CONCLUDED THAT "ST. ANTHONY'S FIRE" COULD NOT BE AN INFECTIOUS DISEASE, GIVEN THAT IT WAS RARE IN CROWDED CITIES BUT COMMON IN RURAL FARMING COMMUNITIES.
FROM FURTHER INVESTIGATION, DR. THUILLER DEDUCED THAT THE ANSWER LAY IN A VICTIM'S DIET.
IN THOSE DAYS, RYE BREAD WAS A STAPLE FOOD FOR POOR COMMUNITIES. THIS WAS DR. THUILLER'S ANSWER.
DR. THUILLER EXAMINED ERGOT OF RYE: SMALL SPURS THAT OCCUR IN THE GRAIN HEAD OF THE RYE PLANT. ALCHEMISTS AND MIDWIVES HAD LONG USED THEM IN MEDICINE, BUT NO ONE HAD EVER CONSIDERED THEM HARMFUL.
ALTHOUGH HE SUSPECTED ERGOT, IT WOULD WOULD BE ANOTHER 200 YEARS BEFORE ANYONE WOULD CONFIRM THIS.

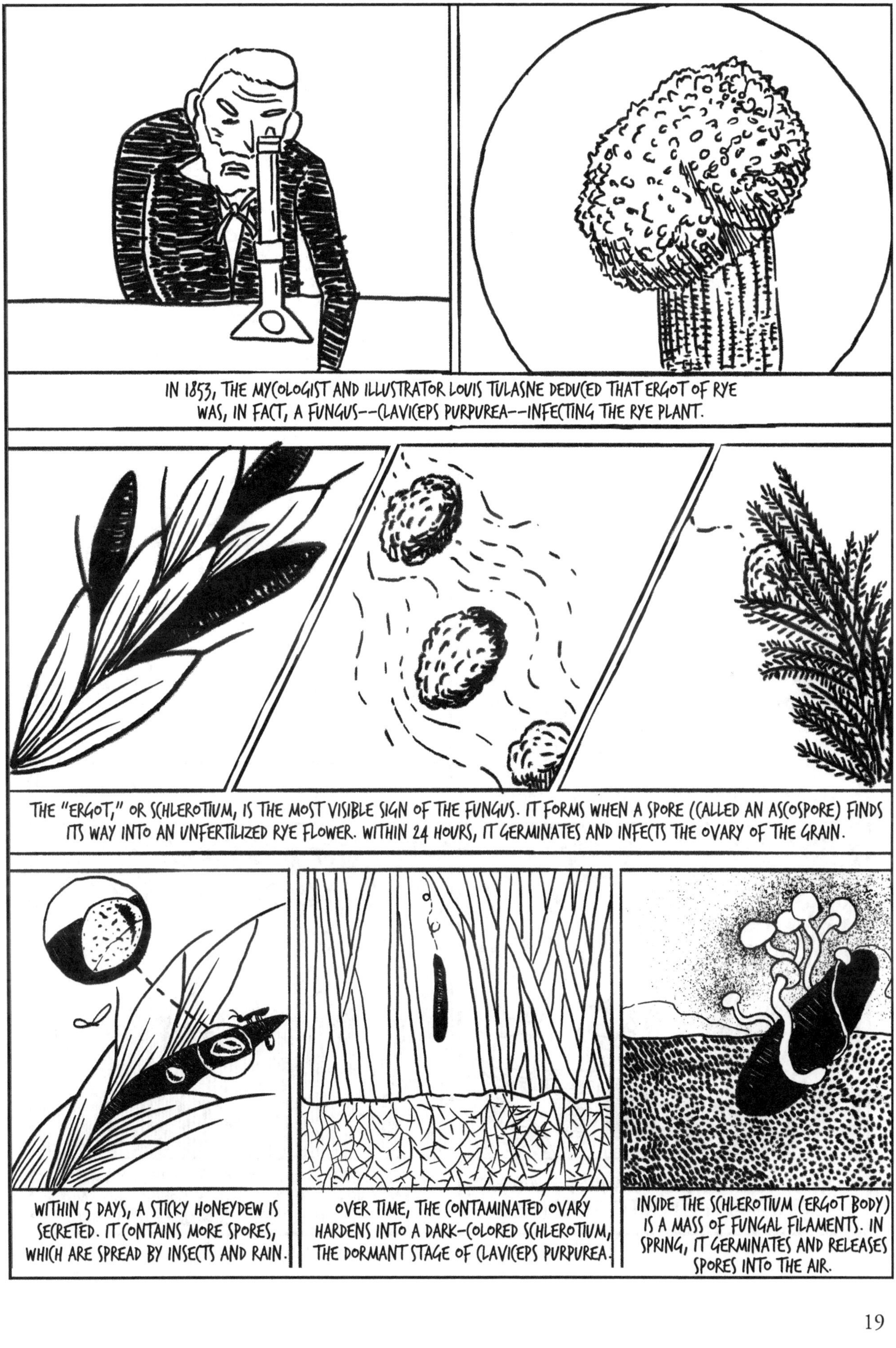
IN 1853, THE MYCOLOGIST AND ILLUSTRATOR LOUIS TULASNE DEDUCED THAT ERGOT OF RYE WAS, IN FACT, A FUNGUS--CLAVICEPS PURPUREA--INFECTING THE RYE PLANT.
THE "ERGOT," OR SCHLEROTIUM, IS THE MOST VISIBLE SIGN OF THE FUNGUS. IT FORMS WHEN A SPORE (CALLED AN ASCOSPORE) FINDS ITS WAY INTO AN UNFERTILIZED RYE FLOWER. WITHIN 24 HOURS, IT GERMINATES AND INFECTS THE OVARY OF THE GRAIN.
WITHIN 5 DAYS, A STICKY HONEYDEW IS SECRETED. IT CONTAINS MORE SPORES, WHICH ARE SPREAD BY INSECTS AND RAIN.
OVER TIME, THE CONTAMINATED OVARY HARDENS INTO A DARK-COLORED SCHLEROTIUM, THE DORMANT STAGE OF CLAVICEPS PURPUREA.
INSIDE THE SCHLEROTIUM (ERGOT BODY) IS A MASS OF FUNGAL FILAMENTS. IN SPRING, IT GERMINATES AND RELEASES SPORES INTO THE AIR.

Roxbury
Dorchester
Boston
Brentry
Crown
Woburn:
Bradford
Alyn
Salem
NEW ENGLAND 1692
OVER TIME, ERGOT POISONING (ERGOTISM) HAS BEEN RESPONSIBLE FOR MORE THAN OUTBREAKS OF "HOLY FIRE." ALL OVER EUROPE, PEOPLE EXHIBITING THE STRANGE SYMPTOMS WERE OFTEN PROSECUTED FOR WITCHCRAFT. AND ACROSS THE SEA, THE MOST FAMOUS WITCH HUNT OF ALL TIME TOOK PLACE IN SALEM, MASSACHUSETTS, 1692.
IN JANUARY, SEVERAL YOUNG GIRLS BEGAN HALLUCINATING AND CONVULSING. THE SYMPTOMS BECAME SO SEVERE THAT WITCHCRAFT WAS BLAMED.
ASKED TO TESTIFY, THE GIRLS BEGAN GIVING NAMES...
...UNTIL 150 PEOPLE HAD BEEN IMPRISONED FOR WITCHCRAFT, WITH 19 CHOOSING TO DIE RATHER THAN CONFESS.
MODERN RESEARCH HAS INDICATED THAT THE GIRLS WERE PROBABLY SUFFERING FROM ERGOT POISONING; THE RAINY SPRING OF 1692 WOULD HAVE ENCOURAGED THE FUNGUS.

FRANCE
MEDITTERRANEAN
PONT-ST.-ESPRIT
THE MOST RECENT CASE OF MASS ERGOT POISONING OCCURRED IN PONT-ST.-ESPRIT, IN THE SOUTH OF FRANCE.
ON AUGUST 15, 1951, THE TOWN HOSPITAL ADMITTED THE FIRST FEW PATIENTS, WHO COMPLAINED OF PAIN IN THE GUT.
WITHIN DAYS, THERE WERE HUNDREDS MORE PATIENTS, AND SYMPTOMS HAD WORSENED TO INCLUDE MUSCULAR SPASM, HALLUCINATIONS, AND FAINTING.
30 malades*
sur les 150 intoxiqués par le mauvais pain de Pont-St-Esprit
deviennent fous
et doivent être internés
* "30 PATIENTS OUT OF THE 150 INTOXICATED BY THE BAD BREAD OF PONT-ST.-ESPRIT HAVE BECOME DERANGED AND MUST BE COMMITED."
ROCH BRIAND
SOME VICTIMS HAD TO BE PUT IN STRAITJACKETS OR TIED TO BEDS TO AVOID INJURING THEMSELVES. THERE ARE EVEN REPORTS OF PEOPLE JUMPING OUT OF WINDOWS.
EVENTUALLY, THE BREAD OF THE LOCAL BAKERY, ROCH BRIAND, WAS FOUND TO CONTAIN TOXIC ALKALOIDS ORIGINATING FROM ERGOT. HOWEVER, IN 2009, AN AMERICAN INVESTIGATIVE JOURNALIST UNEARTHED DOCUMENTS POINTING TO SECRET EXPERIMENTS CONDUCTED BY THE CIA, LEADING HIM TO BELIEVE THAT THE GOVERNMENT AGENCY HAD ACTUALLY CAUSED THE EVENTS IN PONT-ST.-ESPRIT BY DOSING THE TOWN'S BREAD WITH LSD (A HALLUCINOGENIC DRUG).

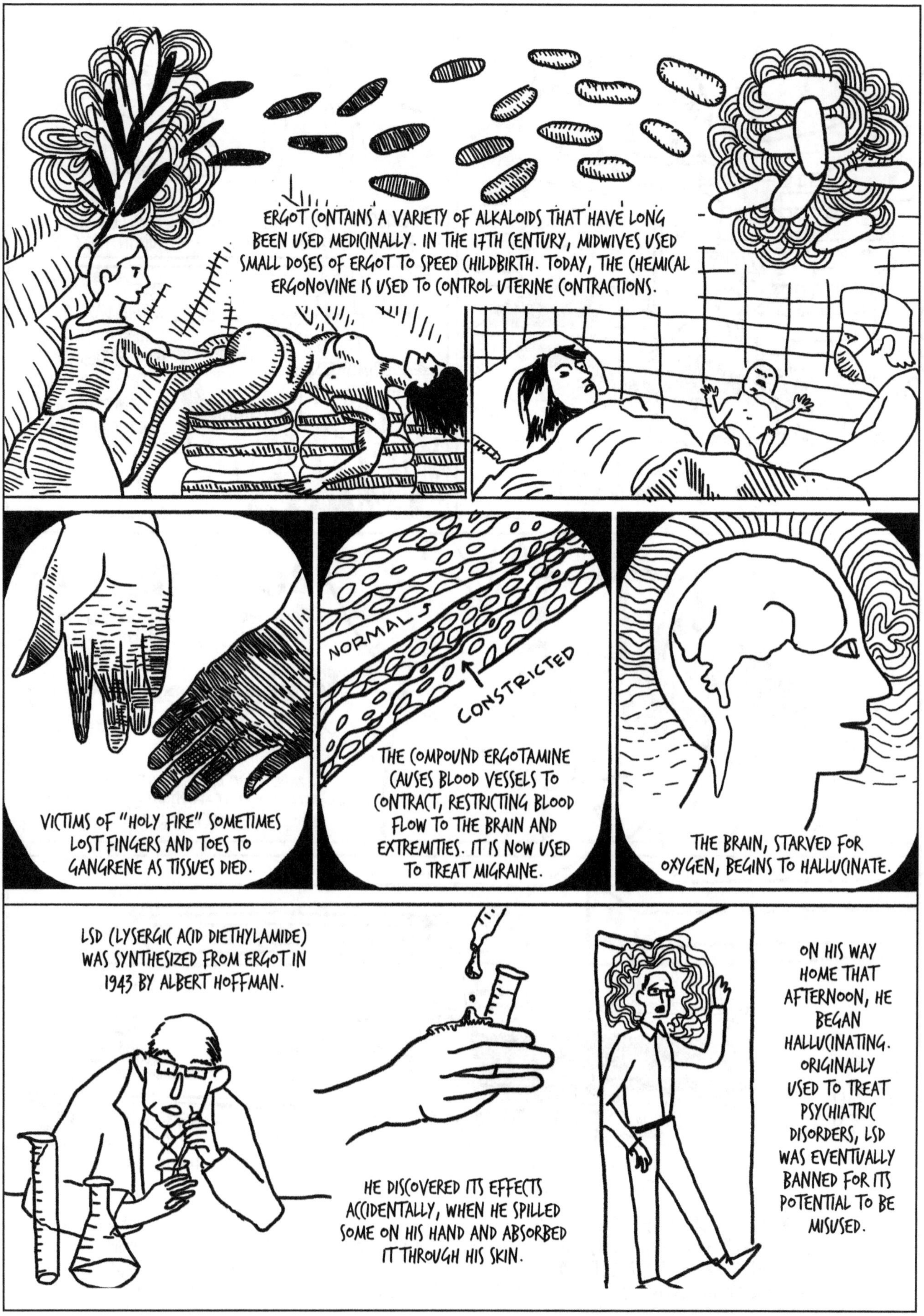
ERGOT CONTAINS A VARIETY OF ALKALOIDS THAT HAVE LONG BEEN USED MEDICINALLY. IN THE 17TH CENTURY, MIDWIVES USED SMALL DOSES OF ERGOT TO SPEED CHILDBIRTH. TODAY, THE CHEMICAL ERGONOVINE IS USED TO CONTROL UTERINE CONTRACTIONS.
VICTIMS OF "HOLY FIRE" SOMETIMES LOST FINGERS AND TOES TO GANGRENE AS TISSUES DIED.
NORMAL
CONSTRICTED
THE COMPOUND ERGOTAMINE CAUSES BLOOD VESSELS TO CONTRACT, RESTRICTING BLOOD FLOW TO THE BRAIN AND EXTREMITIES. IT IS NOW USED TO TREAT MIGRAINE.
THE BRAIN, STARVED FOR OXYGEN, BEGINS TO HALLUCINATE.
LSD (LYSERGIC ACID DIETHYLAMIDE) WAS SYNTHESIZED FROM ERGOT IN 1943 BY ALBERT HOFFMAN.
HE DISCOVERED ITS EFFECTS ACCIDENTALLY, WHEN HE SPILLED SOME ON HIS HAND AND ABSORBED IT THROUGH HIS SKIN.
ON HIS WAY HOME THAT AFTERNOON, HE BEGAN HALLUCINATING. ORIGINALLY USED TO TREAT PSYCHIATRIC DISORDERS, LSD WAS EVENTUALLY BANNED FOR ITS POTENTIAL TO BE MISUSED.

I'LL BET YOU NEVER LEARNED THAT IN HISTORY!
TODAY, BETTER FARMING PRACTICES PREVENTS ERGOTISM. BUT IT WASN'T ALWAYS THAT WAY!

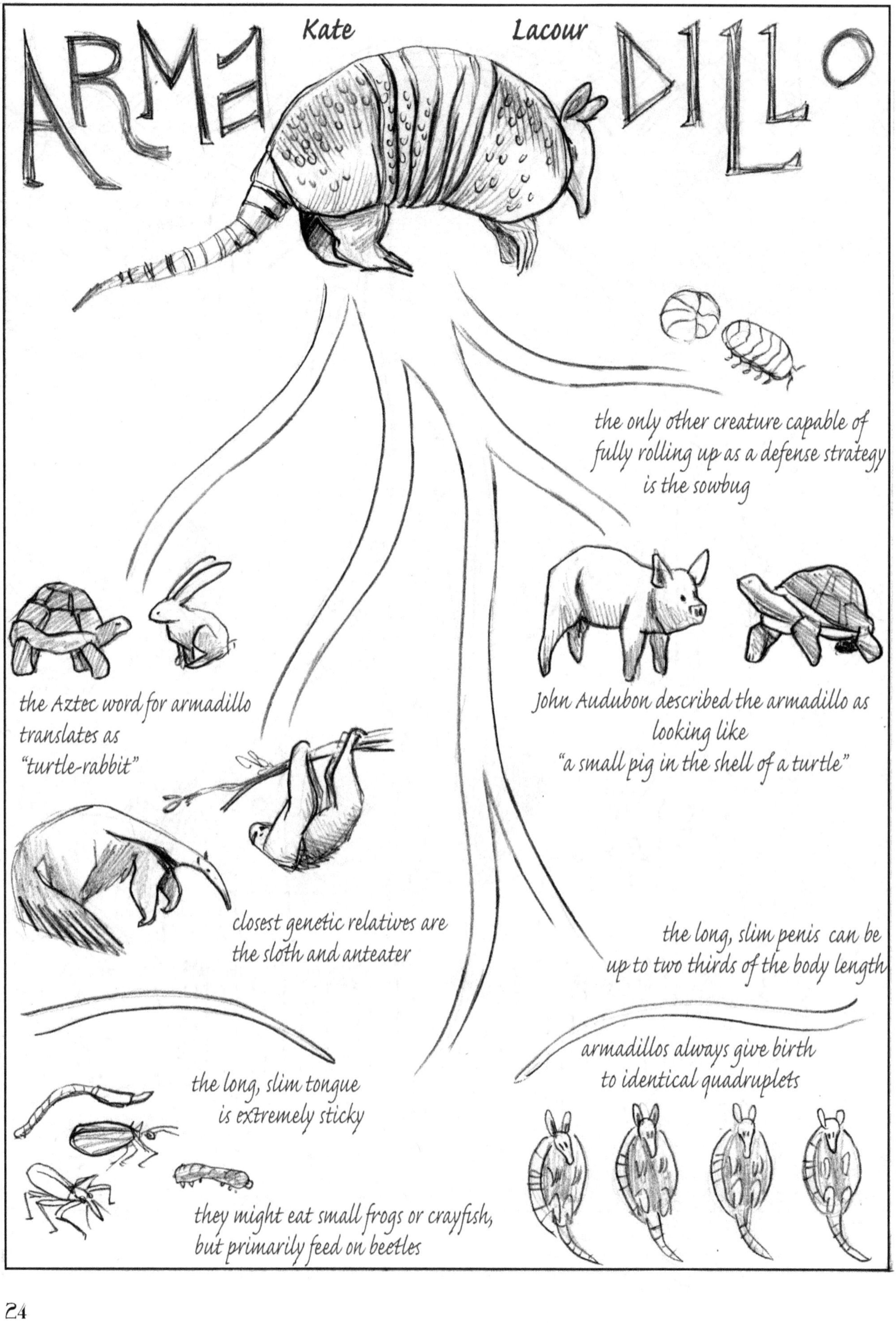
ARMA
Kate
Lacour
DILLO
the only other creature capable of fully rolling up as a defense strategy is the sowbug
the Aztec word for armadillo translates as "turtle-rabbit"
John Audubon described the armadillo as looking like "a small pig in the shell of a turtle"
closest genetic relatives are the sloth and anteater
the long, slim penis can be up to two thirds of the body length
armadillos always give birth to identical quadruplets
the long, slim tongue is extremely sticky
they might eat small frogs or crayfish, but primarily feed on beetles

Aphid-farming Ants
By Megan Archer
If you see lots of ants running up and down a plant...
...they may be farming aphids!

Dairying ants obtain nutrients by raising and protecting a herd of aphids in order to eat their sugary secretions, in a relationship reminiscent of human dairy farmers and their cows.

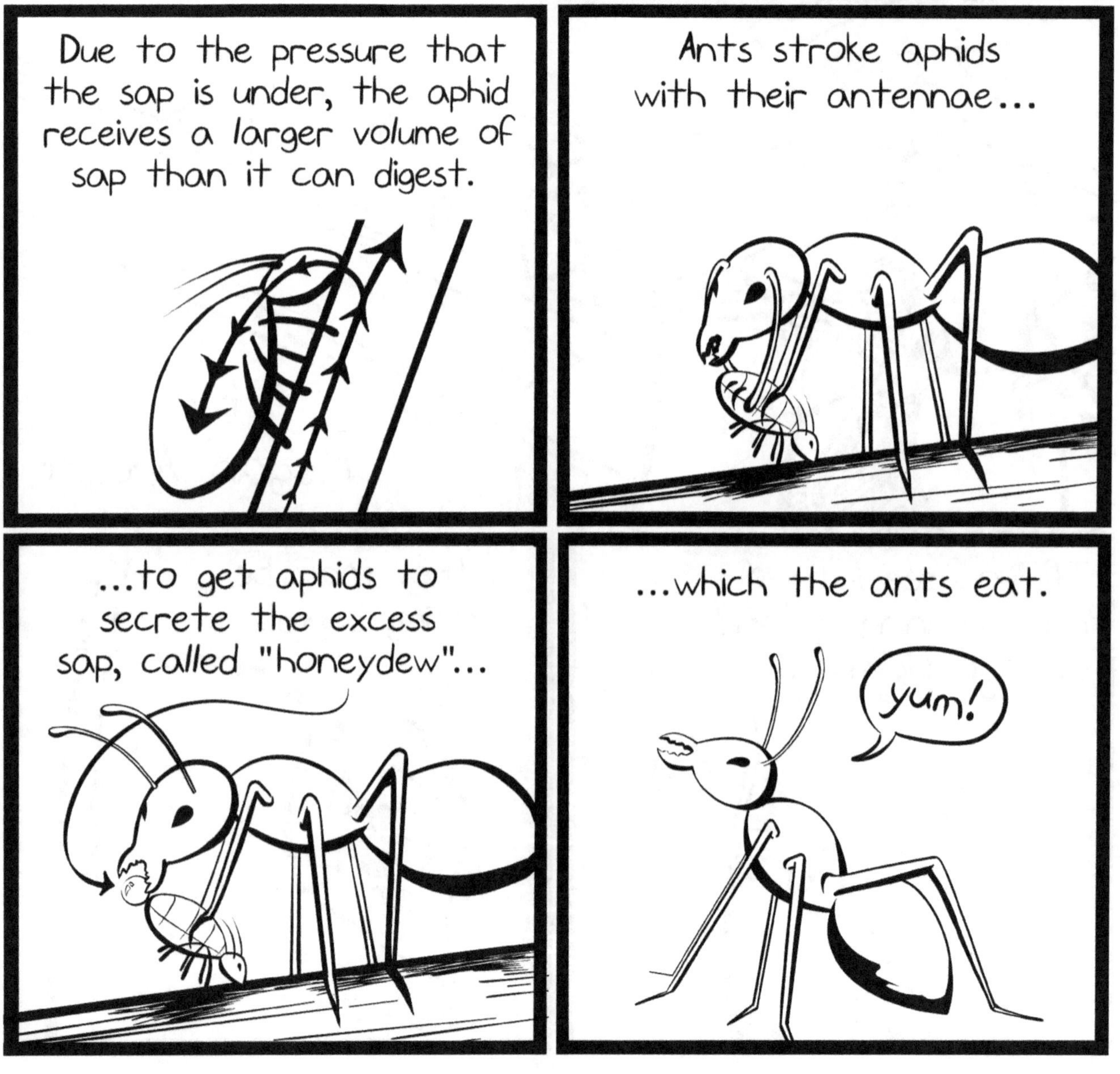

So how is the life of a farmed aphid?
?
PROS
Ants protect their aphids.
They fight off the aphid's predators,
and carry their aphids to shelter when it rains.
Ants even help ensure aphid eggs will hatch in the spring
by keeping them safe in their nest over the winter.

—CONS—

In exchange for protection, the aphids are trapped and controlled by the ants.
To keep aphids from escaping, ants chew off their wings.
New research suggests that ants may even keep aphids docile with chemicals released from their feet.
? ?
Ants have also been observed eating some of their aphids.
But overall...
not a bad life for an aphid!

"I picked away at the layers of slime to reveal the most beautiful fish I had ever seen. It was five feet long, a pale mauvy blue with faint flecks of whitish spots; it had an iridescent silver-blue-green sheen all over. It was covered in hard scales, and it had four limb-like fins and a strange little puppy dog tail.
It was such a beautiful fish - more like a big china ornament - but I didn't know what it was."
In 1938, three days before Christmas, in the town of East London on the cape of South Africa, a fishing boat trawled up an unusual find.
This creature would later be identified as a *living fossil*— a term coined by Charles Darwin to describe an animal or plant closely resembling a species thought to have gone extinct, its existence heretofore known only in the fossil record.
A woman named Marjorie Courtenay-Latimer was the first to recognize the fish as something truly extraordinary.
DON'T CALL IT A COMEBACK
by hedj.

Courtenay-Latimer was born on February 24, 1907, over two months prematurely. As a child, she was frequently ill, but her health didn't keep her from becoming a passionate naturalist. At the age of 24, she became the sole curator of the East London Museum.
Finding it an "empty shell," she burned the handful of bug-infested specimens she found in the basement and replaced them with items from her own collection. For over forty years, she would devote herself to the museum's upkeep, gathering items for display on the weekends and inspecting the nets of local fishermen for unusual specimens.
On December 22, 1938, she received a call to examine the catch of Captain Hendrik Goosen, whom she had befriended some years previously on a trip to Bird Island. It was then that she made a startling discovery.
Courtenay-Latimer recognized its scales and four limb-like fins as being characteristic of *ganoid*, or fossil fish, and couldn't find anything resembling it in any of her reference books. The local mortuary and cold storage both refused to store the fish, so she wrapped it in formaldehyde and telephoned James Leonard Brierley Smith, a chemistry lecturer and amateur ichthyologist. Unable to reach him, she sent a letter.
EAST LONDON MUSEUM
23rd Dec, 1938
Dear Dr. Smith,
I had the most queer looking specimen brought to my attention yesterday, the Capt. of The trawler told me about it so I immediately set off to see the specimen which I had removed to our Taxidermist as soon as I could. I however have drawn a rough sketch and am in hopes that

A response from JLB Smith would not arrive for another 11 days. When he finally read the letter, he identified the fish, from only a description and crude drawing, as resembling a fossil fish he had seen in a scientific journal, found in Greenland and measuring 12 inches long— and dated at 200 million years old.

By the time Smith arrived in East London to see the fish in person, its innards had already been disposed of. Courtenay-Latimer and the taxidermist had found the flesh of the fish to be pure white and "clay-like"; it had no ribs, and a flexible oil-filled tube rather than a spine. Inspecting the fish, Smith confirmed his suspicions that it was, in fact, a *coelacanth*— a lobe-finned fish of an order thought to have gone extinct over 65 million years ago.

Reactions to the fish's discovery were immediate and varied. Newspapers sensationalized the coelacanth as an evolutionary "missing link," which outraged religious fundamentalists; meanwhile, museums and scientific institutions began searching for more specimens to dissect. Some criticized Smith for naming the extant species *Latimeria chalumnae* after Courtenay-Latimer, believing her responsible for the loss of the first fish's viscera, but Smith's response was firm:

"It was the energy and determination of Miss Latimer which saved so much, and scientific workers have good cause to be grateful. The genus *Latimeria* stands as my tribute."

Since 1938, almost 200 more coelacanths have been caught, and as technology has advanced, biologists have been able to observe these fish in their natural habitat. Coelacanths have never been held in captivity, and their habitat is not suited to eco-tourism, so most people will never see a live coelacanth in the flesh. Conservation efforts have been undertaken out of concern for the fish's continued survival, but because of its reclusive nature, there is no sure way of knowing the status of its population.

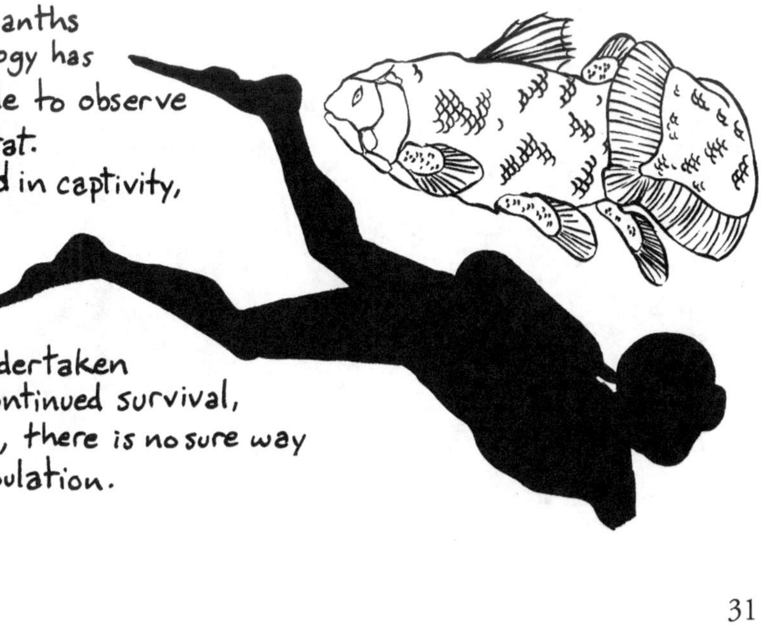

Modern coelacanths live at depths of over 100m, in various parts of the Indian Ocean. The coelacanth is a surprisingly graceful swimmer and has been observed swimming on its back as well as on its head, in a position known as a "headstand." It rests in caves during the day, in groups or singly, emerging at night to feed alone. Unlike most other fish, the coelacanth's swim bladder is filled with oil, not air. Also unlike other fish, its blood is similar to a shark's, with a high quantity of urea to maintain osmotic balance—that is, to stay hydrated in seawater. The coelacanth's metabolism is one of the slowest of any known fish; it's a slow mover and can go for long periods without feeding.

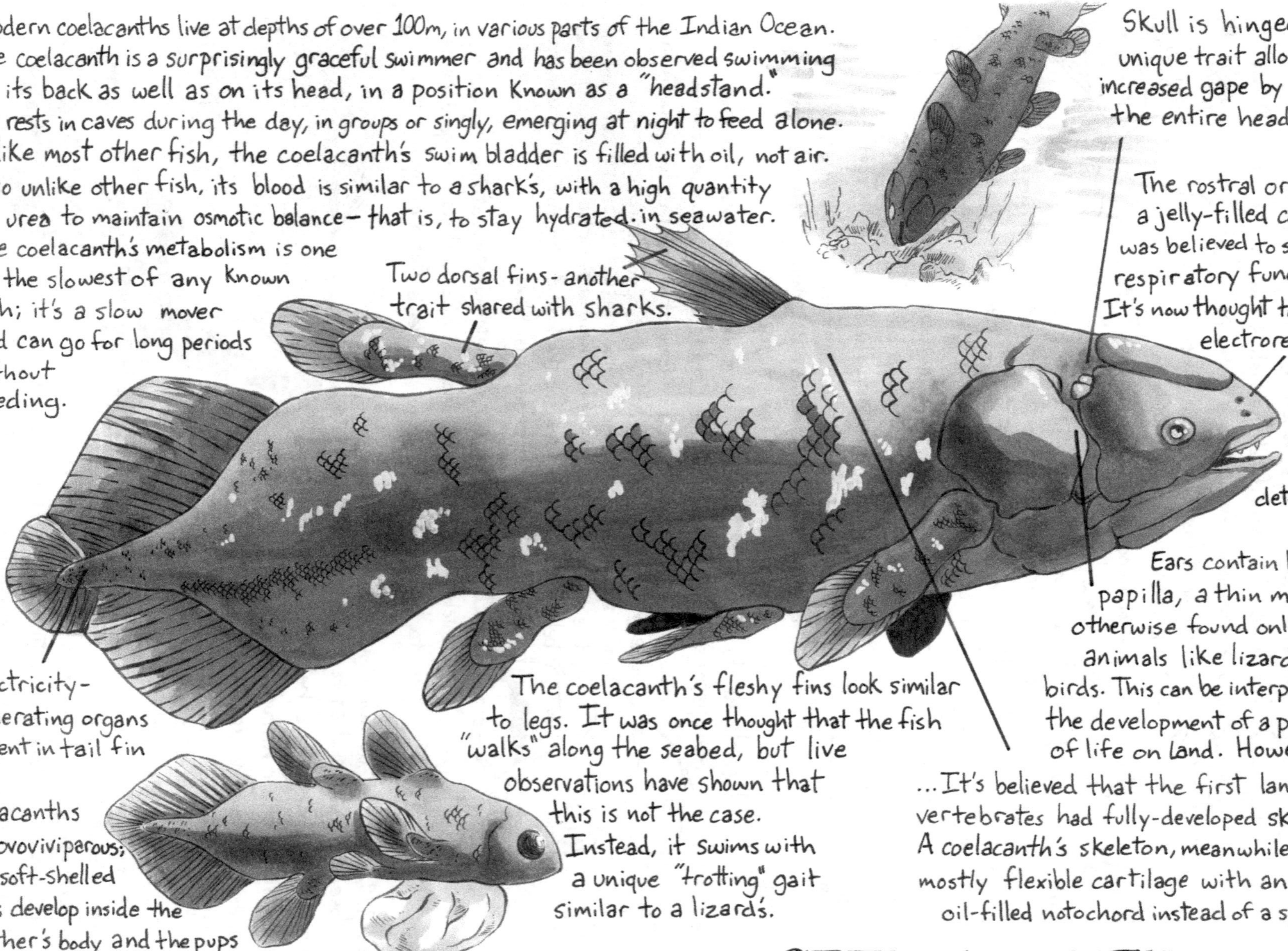

Skull is hinged—a unique trait allowing an increased gape by opening the entire head

The rostral organ, a jelly-filled cavity, was believed to serve a respiratory function. It's now thought to be an electroreceptive organ, used to navigate and to detect prey.

Two dorsal fins- another trait shared with sharks.

Ears contain basilar papilla, a thin membrane otherwise found only in land animals like lizards and birds. This can be interpreted as the development of a prerequisite of life on land. However...

...It's believed that the first land vertebrates had fully-developed skeletons. A coelacanth's skeleton, meanwhile, is mostly flexible cartilage with an oil-filled notochord instead of a spine.

Electricity-generating organs present in tail fin

The coelacanth's fleshy fins look similar to legs. It was once thought that the fish "walks" along the seabed, but live observations have shown that this is not the case. Instead, it swims with a unique "trotting" gait similar to a lizard's.

Coelacanths are ovoviviparous; the soft-shelled eggs develop inside the mother's body and the pups are born live. One female was found with 26 pups inside her uterus.

SEEL-uh-CANTH

SEPIA. BY ETHAN KOCAK
I'VE WANTED TO KEEP CUTTLEFISH FOR ALMOST AS LONG AS I'VE BEEN AN AQUARIST. WHEN I WAS A TEENAGER, I SUCCESSFULLY LEARNED ENOUGH ABOUT MARINE AQUARIA TO KEEP A FEW OCTOPUSES, BUT CUTTLEFISH IN THE UNITED STAES WERE HARD TO COME BY. BUT NOW, IN MY THIRTIES, AND THANKS TO THE INTERNET, I WAS EASILY ABLE TO FIND *SEPIA BANDENSIS* EGGS, THE DWARF CUTTLEFISH, FOR SALE. I MANAGED TO RAISE THEM FROM EGG, TO DEATH, A PROCESS WHICH SADLY TOOK ONLY ABOUT 8 MONTHS.
MANY CEPHALOPODS ARE VERY SHORT-LIVED.
SO MAYBE IT WAS THIS MEDITATION ON AQUARIA AND MORTALITY THAT HAD ME TURNING MY THOUGHTS TO THE EFFORTS OF THOSE WHO HAD COME BEFORE ME. WHO WERE THEY? HOW DID WE ARRIVE AT THIS CUSTOM OF KEEPING GLASS BOXES WITH LITTLE ALIEN WORLDS IN THEM?

WHEN YOU START RESEARCHING THE HISTORY OF AQUARIUMS, THE FIRST PERSON YOU TEND TO COME ACROSS IS A MAN BY THE NAME OF PHILIP HENRY GOSSE. PHILIP WAS A VERY POPULAR SCIENTIST AND SCIENCE COMMUNICATOR OF HIS DAY.
IN 1859, HE PUBLISHED "A HANDBOOK TO THE MARINE AQUARIUM" WHICH DETAILED HOW TO SETUP AND MAINTAIN AQUARIA WHERE VICTORIANS COULD MAINTAIN ANIMALS THEY HAD COLLECTED FROM TIDE POOLS AND SHORELINES.
IT WAS PHILIP WHO COINED THE WORD "AQUARIUM" RATHER THAN THE UNWIELDY "AQUATIC VIVARIUM," WHICH HAD BEEN IN USE BEFORE.

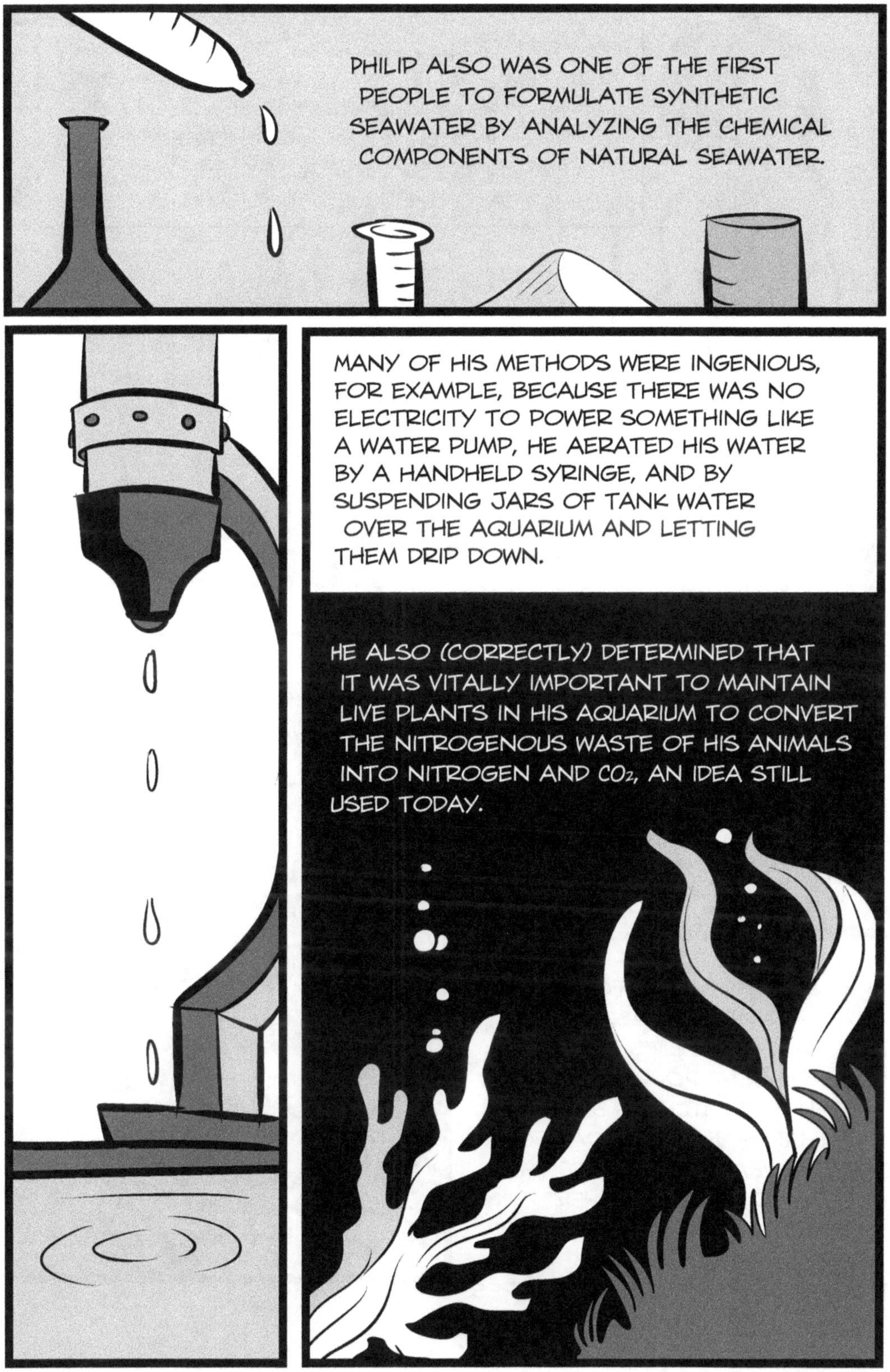
PHILIP ALSO WAS ONE OF THE FIRST PEOPLE TO FORMULATE SYNTHETIC SEAWATER BY ANALYZING THE CHEMICAL COMPONENTS OF NATURAL SEAWATER.
MANY OF HIS METHODS WERE INGENIOUS, FOR EXAMPLE, BECAUSE THERE WAS NO ELECTRICITY TO POWER SOMETHING LIKE A WATER PUMP, HE AERATED HIS WATER BY A HANDHELD SYRINGE, AND BY SUSPENDING JARS OF TANK WATER OVER THE AQUARIUM AND LETTING THEM DRIP DOWN.
HE ALSO (CORRECTLY) DETERMINED THAT IT WAS VITALLY IMPORTANT TO MAINTAIN LIVE PLANTS IN HIS AQUARIUM TO CONVERT THE NITROGENOUS WASTE OF HIS ANIMALS INTO NITROGEN AND CO_2, AN IDEA STILL USED TODAY.

A CONTEMPORARY OF PHILIP HENRY GOSSE WAS THE MOST SKILLED AQUARIST OF THE VICTORIAN AQUARIUM FAD. HER NAME WAS ANNA THYNNE. ANNA WAS MARRIED TO THE SUB-DEAN OF WESTMINSTER ABBEY AND THEREFORE HAPPENED TO BE WEALTHY. SHE WAS A MOTHER OF 10 CHILDREN.
IN 1846, SHE BUILT A MARINE AQUARIUM AND STOCKED IT WITH SPONGES AND CORALS NATIVE TO COASTAL ENGLAND.
SHE KEPT THEM ALIVE FOR OVER THREE YEARS (AT LEAST A YEAR LONGER THAN PHLIP'S PERSONAL AQUARIA). SHE POURED TANK WATER INTO PITCHERS BACK AND FORTH BY HAND EVERY DAY FOR THOSE 3 YEARS TO AERATE THE WATER AND OFTEN HAD HER SERVANTS DO THIS AS WELL, WHO WERE AMUSED BY ANNA'S FASCINATION WITH SEA CREATURES.

1847
SHE DISPLAYED HER AQUARIA IN THE CLOISTERS OF THE ABBEY ITSELF, AND IT BROUGHT NATURALISTS FROM ALL OVER. IT WAS THE FIRST PUBLIC AQUARIUM IN ENGLAND.
BUT WHAT OF THE VERY FIRST AQUARIUM?

OF COURSE, HUMANS HAVE KEPT FISH LONGER THAN THEY'VE KEPT RECORDS.
MONKS RAISED CARP IN PONDS IN THE MIDDLE AGES FOR FOOD,
ROMANS BEFORE THEM USED THEIR EXPERTISE WITH AQUEDUCTS TO CRAFT CONTAINERS FOR FISH.
ARTISANS IN CHINA MADE BEAUTIFUL PORCELAIN BOWLS SPECIFICALLY FOR THE KEEPING AND BREEDING OF GOLDFISH, WHICH THEY DOMESTICATED FROM ASIATIC CARP.
BUT WHO MADE THE FIRST RECOGNIZABLY MODERN AQUARIUM?
A WOMAN NAMED JEANNE VILLEPREUX-POWER.

BEFORE SHE BECAME A SCIENTIST,
SHE WAS A DRESSMAKER FOR THE FRENCH ROYALTY.
SHE MARRIED A MERCHANT BY THE NAME OF JAMES POWER.

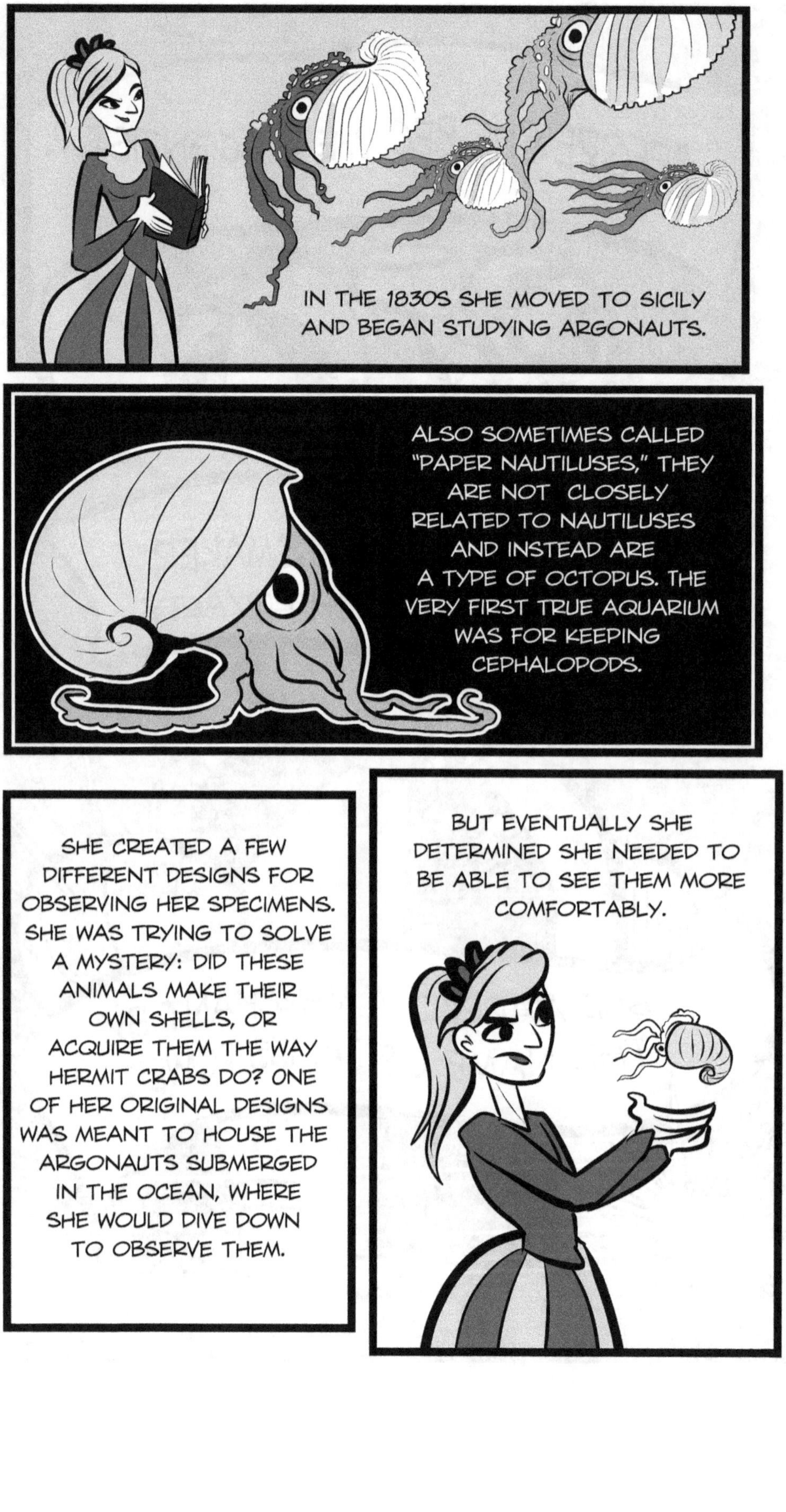
IN THE 1830S SHE MOVED TO SICILY AND BEGAN STUDYING ARGONAUTS.
ALSO SOMETIMES CALLED "PAPER NAUTILUSES," THEY ARE NOT CLOSELY RELATED TO NAUTILUSES AND INSTEAD ARE A TYPE OF OCTOPUS. THE VERY FIRST TRUE AQUARIUM WAS FOR KEEPING CEPHALOPODS.
SHE CREATED A FEW DIFFERENT DESIGNS FOR OBSERVING HER SPECIMENS. SHE WAS TRYING TO SOLVE A MYSTERY: DID THESE ANIMALS MAKE THEIR OWN SHELLS, OR ACQUIRE THEM THE WAY HERMIT CRABS DO? ONE OF HER ORIGINAL DESIGNS WAS MEANT TO HOUSE THE ARGONAUTS SUBMERGED IN THE OCEAN, WHERE SHE WOULD DIVE DOWN TO OBSERVE THEM.
BUT EVENTUALLY SHE DETERMINED SHE NEEDED TO BE ABLE TO SEE THEM MORE COMFORTABLY.

JEANNE ESTABLISHED, CORRECTLY, THAT FEMALE ARGONAUTS SECRETE THEIR OWN SHELLS.
BY RAISING HER ANIMALS IN CAPTIVITY, SHE REALIZED THAT MALE ARGONAUTS LACKED THE SHELL (AND THEREFORE HAD NOT BEEN IDENTIFIED AS ARGONAUTS UP TO THAT POINT).
AND THAT NEITHER USED THEIR MODIFIED ARMS FOR ROWING AS WAS COMMONLY BELIEVED.
WHEEEEEEEE
SHE WAS SO TALENTED AS A SCIENTIST THAT EVEN RICHARD OWEN, THE FAMOUS ZOOLOGIST AND ANATOMIST, WROTE ABOUT HER AS THE "MOTHER OF AQUARIOPHILY"--EVEN THOUGH SHE WAS ABOUT 30 YEARS AHEAD OF THE HUGE BOOM IN PERSONAL AQUARIA.

SADLY, THERE'S A REASON JEANNE VILLEPREUX-POWER IS MOSTLY FORGOTTEN TODAY.
WHEN SHE AND JAMES DECIDED TO LEAVE SICILY, ALMOST ALL OF JEANNE'S PAPERS, ILLUSTRATIONS, AND RESEARCH WERE LOST IN A SHIPWRECK. ALTHOUGH SHE WAS NOT TRAVELING ON THAT SHIP, SHE NEVER RESUMED HER SCIENTIFIC PURSUITS.
HOWEVER, IN 1997, A CRATER WAS NAMED AFTER JEANNE ON VENUS AS A WAY TO COMMEMORATE HER GREAT ACHIEVEMENTS.

PHILIP HENRY GOSSE
ANNA THYNNE
JEANNE VILLEPREUX-POWER

BECAUSE OF THEM, WE ARE ABLE TO HAVE THE PRIVILEGE OF LITTLE ALIEN WORLDS IN GLASS BOXES, RIGHT IN OUR HOMES.

TO HAVE WHAT THE VICTORIANS CALLED "PARLOUR OCEANS" IS A GIFT, AND WE SHOULD REMEMBER THE SCIENTISTS THAT MADE IT POSSIBLE.

EASTER SUNDAY, IN A BACKYARD IN WEST VIRGINIA:
A MOMENT OF SILENCE FOR BUD, THE GRAND OLD HORNED FROG.* THE LAST SURVIVOR OF THE PETS FROM MY YOUTH. THE END OF AN ERA.
* CERATOPHRYS ORNATA
RESURRECTION
THE REAL-LIFE SUPERPOWERS OF FROGS AND MOSSES
BY ROSS WOOD STUDLAR
EDITED BY SUSAN MOYLE STUDLAR
FONT DESIGNED BY ADAM STAFFARONI
I NEVER IMAGINED THAT THE AMPHIBIAN WOULD MAKE IT TO SEE MY 34TH BIRTHDAY, AND PERHAPS HIS OWN 20TH.
HE MOVED.
WHAT?
HE MOVED.
CARL IS RIGHT. I SAW IT, TOO.
OH JOY OF JOYS— BUD IS ALIVE!
HE WAS JUST COMATOSE FOR A LITTLE WHILE.
GOOD WORK, CARL AND DAD. YOUR KEEN EYES SPARED HIM A PREMATURE BURIAL.
AMPHIBIANS (SUCH AS FROGS AND TOADS) HAVE EXTRAORDINARY POWERS, TO APPARENTLY DIE AND RETURN TO LIFE.
EVER WONDER WHAT HAPPENS TO A COLD-BLOODED FROG ON A FREEZING WINTER DAY?
I RECALL READING ABOUT IT IN LIFE IN THE COLD.
I REMEMBER GIVING YOU THAT BOOK. AS A KID GROWING UP IN MINNESOTA, I LOVED WINTER BUT WONDERED HOW PLANTS AND ANIMALS SURVIVED THE COLD (UP TO 30 DEGREES BELOW ZERO FAHRENHEIT)!
ARCTIC EXPLORERS RECKONED WITH THAT MYSTERY TOO, MOM!

CAPTAIN FRANCIS SMITH LANDED IN THE CANADIAN ARCTIC IN MAY 1747. HE HEARD WOOD FROGS (RANA SYLVATICA) CALLING FROM A MELTWATER POND. YET THEY HAD BEEN BURIED AND FROZEN SOLID ALL WINTER!
RACKET RACKET
RACKET RACKET
WHEN THE WEATHER TURNED COLD AGAIN, THE FROGS REFROZE.
AND SMITH PERFORMED WHAT HE CALLED "A REMARKABLE EXPERIMENT."
"TO TAKE THE EARTH IN WHICH THE FROG IS SO FROZE, AND TO BREAK THAT EARTH IN PIECES WITHOUT THAWING IT, THE FROG WILL THEN BREAK WITH IT AS SHORT AS A PIECE OF GLASS."
KRUNCH!
"BUT...LAY THAT EARTH AT A SMALL DISTANCE FROM THE FIRE, SO AS TO THAW IT, THE FROG WILL RECOVER HIS SUMMER ACTIVITY, AND LEAP AS USUAL."
IT WAS NOT UNTIL 150 YEARS LATER, IN THE 1980S, THAT SCIENTISTS MADE SERIOUS INQUIRY ON HOW FROGS CAN WITHSTAND BEING FROZEN. DR. KENNETH STOREY WAS AMONG THE FIRST TO FREEZE FROGS IN A LABORATORY SETTING, AND MONITOR THEIR PHYSIOLOGY. IT TURNS OUT THAT WHEN WINTER WEATHER MOVES IN, A LITTLE WOOD FROG UNDERGOES BIG CHANGES.
THUMP-THUMP
THUMP-THUMP

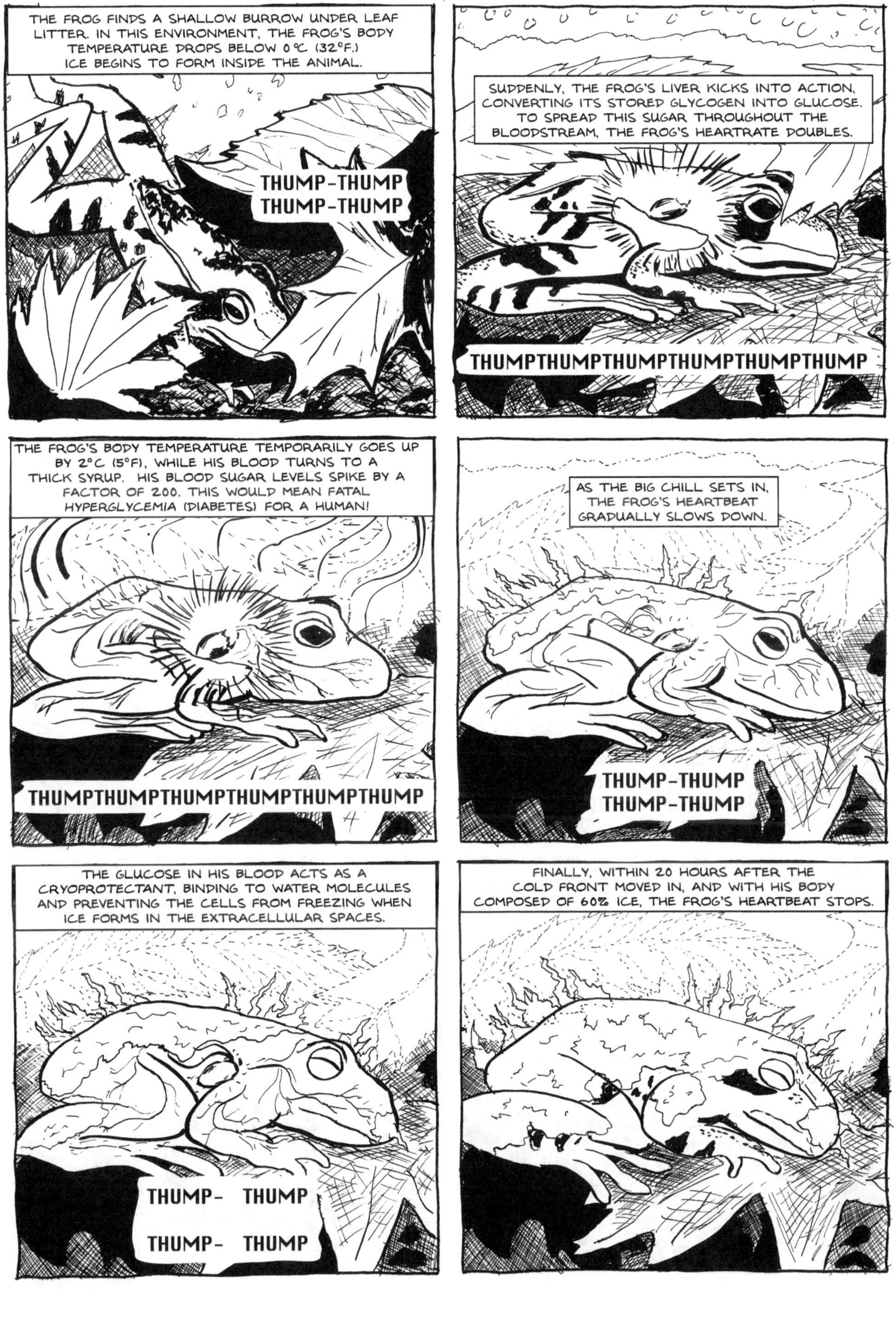
THE FROG FINDS A SHALLOW BURROW UNDER LEAF LITTER. IN THIS ENVIRONMENT, THE FROG'S BODY TEMPERATURE DROPS BELOW 0°C (32°F.) ICE BEGINS TO FORM INSIDE THE ANIMAL.
THUMP-THUMP THUMP-THUMP
SUDDENLY, THE FROG'S LIVER KICKS INTO ACTION, CONVERTING ITS STORED GLYCOGEN INTO GLUCOSE. TO SPREAD THIS SUGAR THROUGHOUT THE BLOODSTREAM, THE FROG'S HEARTRATE DOUBLES.
THUMPTHUMPTHUMPTHUMPTHUMPTHUMP
THE FROG'S BODY TEMPERATURE TEMPORARILY GOES UP BY 2°C (5°F), WHILE HIS BLOOD TURNS TO A THICK SYRUP. HIS BLOOD SUGAR LEVELS SPIKE BY A FACTOR OF 200. THIS WOULD MEAN FATAL HYPERGLYCEMIA (DIABETES) FOR A HUMAN!
THUMPTHUMPTHUMPTHUMPTHUMPTHUMP
AS THE BIG CHILL SETS IN, THE FROG'S HEARTBEAT GRADUALLY SLOWS DOWN.
THUMP-THUMP THUMP-THUMP
THE GLUCOSE IN HIS BLOOD ACTS AS A CRYOPROTECTANT, BINDING TO WATER MOLECULES AND PREVENTING THE CELLS FROM FREEZING WHEN ICE FORMS IN THE EXTRACELLULAR SPACES.
THUMP- THUMP
THUMP- THUMP
FINALLY, WITHIN 20 HOURS AFTER THE COLD FRONT MOVED IN, AND WITH HIS BODY COMPOSED OF 60% ICE, THE FROG'S HEARTBEAT STOPS.

AND YET THE ANIMAL RETAINS THE SPARK OF LIFE. HE IS KEPT GOING BY THE ANAEROBIC METABOLISM OF HIS ENERGY STORES AND THE CRYOPROTECTANT ACTION OF GLUCOSE.
WHEN THE WEATHER WARMS UP, THE FROG STARTS TO THAW. WITHIN AN HOUR, HIS HEART RESUMES BEATING.
THUMP- THUMP
THUMP- THUMP
SIX HOURS LATER, EVEN IF THE FROG HAS ONLY WARMED UP TO 5°C (41°F), HIS HEARTBEAT IS BACK TO NORMAL.
THUMP-THUMP
THUMP-THUMP
AND HE SINGS IN THE STREAM IN HOPES OF ATTRACTING A MATE! FROGS CAN SLIP IN AND OUT OF A FROZEN STATE QUICKLY—AN IMPORTANT ADAPTATION FOR LIFE IN THE SPRING BREEDING SEASON, WHEN TEMPERATURES RISE AND FALL.
RACKET RACKET
RACKET RACKET
IN A WORLD FULL OF COLD PLACES, FREEZE-TOLERANT FROGS ARE FOUND ONLY IN THE AMERICAS.
DR. STOREY CONTENDS THAT FREEZE TOLERANCE EVOLVED FROM DESICCATION TOLERANCE—ANOTHER ESSENTIAL ABILITY IN AMERICA'S VARIED CLIMATE.
IMAGINE THE CHALLENGES THAT OUR DESERTS POSE TO AMPHIBIANS.
SPADEFOOT TOADS (FAMILY SCAPHIOPODIDAE) ARE SUPREMELY ADAPTED TO DESERTS AND CAN LOSE UP TO 50% OF THEIR BODY WATER WITHOUT HARM. THESE REMARKABLE CREATURES STAY QUIETLY BURIED UNDERGROUND FOR UP TO TWO YEARS AT A STRETCH AND EMERGE WITH SPRING RAINS TO GORGE THEMSELVES ON INSECTS AND ENGAGE IN MATING ORGIES.

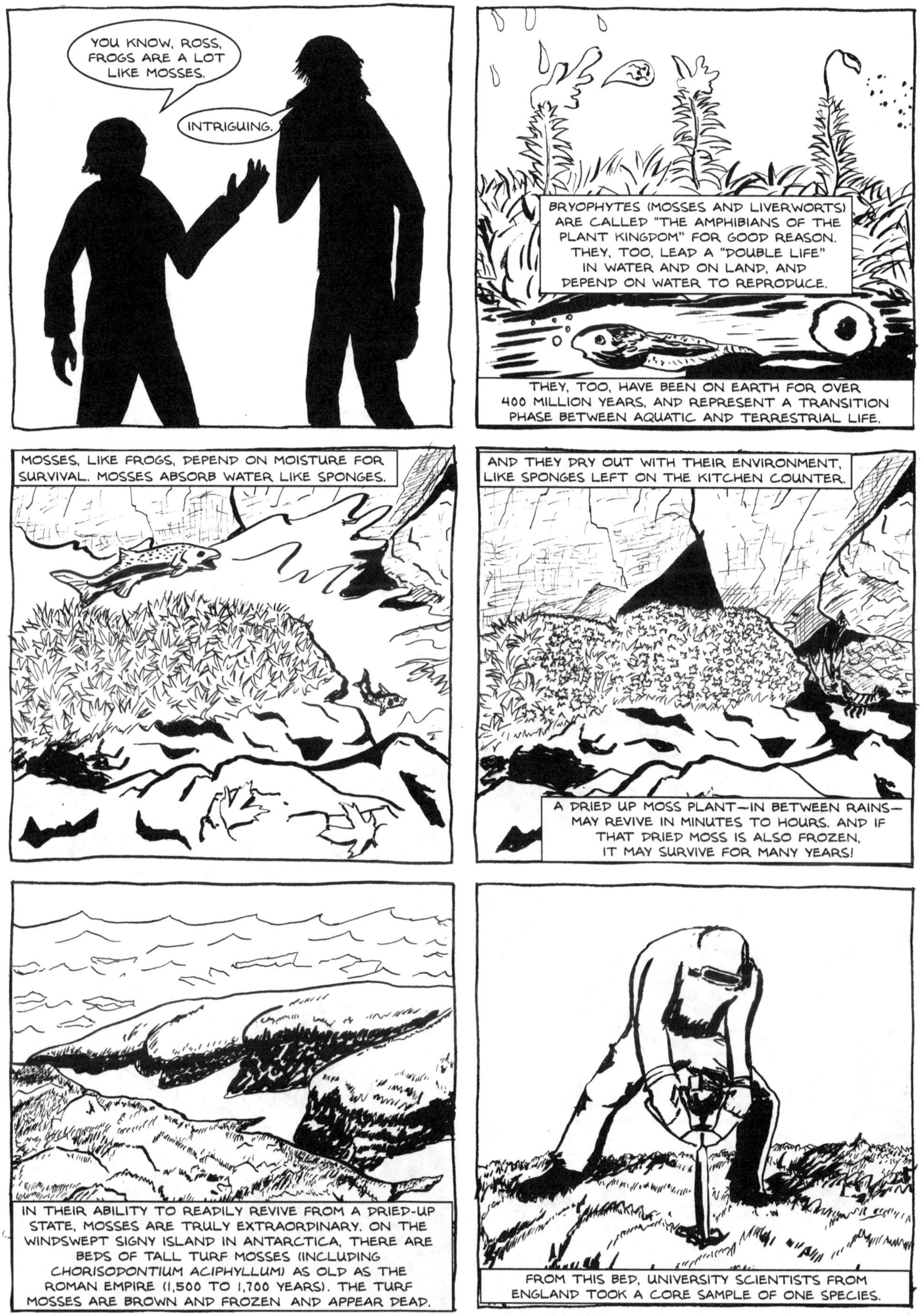
YOU KNOW, ROSS, FROGS ARE A LOT LIKE MOSSES.
INTRIGUING.
BRYOPHYTES (MOSSES AND LIVERWORTS) ARE CALLED "THE AMPHIBIANS OF THE PLANT KINGDOM" FOR GOOD REASON. THEY, TOO, LEAD A "DOUBLE LIFE" IN WATER AND ON LAND, AND DEPEND ON WATER TO REPRODUCE.
THEY, TOO, HAVE BEEN ON EARTH FOR OVER 400 MILLION YEARS, AND REPRESENT A TRANSITION PHASE BETWEEN AQUATIC AND TERRESTRIAL LIFE.
MOSSES, LIKE FROGS, DEPEND ON MOISTURE FOR SURVIVAL. MOSSES ABSORB WATER LIKE SPONGES.
AND THEY DRY OUT WITH THEIR ENVIRONMENT, LIKE SPONGES LEFT ON THE KITCHEN COUNTER.
A DRIED UP MOSS PLANT—IN BETWEEN RAINS—MAY REVIVE IN MINUTES TO HOURS. AND IF THAT DRIED MOSS IS ALSO FROZEN, IT MAY SURVIVE FOR MANY YEARS!
IN THEIR ABILITY TO READILY REVIVE FROM A DRIED-UP STATE, MOSSES ARE TRULY EXTRAORDINARY. ON THE WINDSWEPT SIGNY ISLAND IN ANTARCTICA, THERE ARE BEDS OF TALL TURF MOSSES (INCLUDING CHORISODONTIUM ACIPHYLLUM) AS OLD AS THE ROMAN EMPIRE (1,500 TO 1,700 YEARS). THE TURF MOSSES ARE BROWN AND FROZEN AND APPEAR DEAD.
FROM THIS BED, UNIVERSITY SCIENTISTS FROM ENGLAND TOOK A CORE SAMPLE OF ONE SPECIES.

BACK AT THE LAB, THEY CHOPPED THE CORE INTO SMALLER PIECES, PLACED THEM IN SEALED CONTAINERS, AND EXPOSED THEM TO LIGHT AND TEMPERATURE LEVELS SIMILAR TO THOSE FOUND IN THE MOSS'S HOME.
IN THREE TO SIX WEEKS, THE MOSS SPROUTED NEW GREEN FUZZ! THE SPARK OF LIFE ENDURED IN THE FROZEN MOSS FOR 1,500 YEARS.
ONE YEAR EARLIER, CANADIAN SCIENTISTS HAD SIMILARLY REVIVED FOUR MOSS SPECIES FROZEN UNDER THE ARCTIC TEARDROP GLACIER FOR FOUR CENTURIES. AS ICE MELTS WORLD WIDE, PERHAPS MANY "LONG DEAD" MOSSES WILL WAKE UP AND REJOIN THE WORLD OF THE LIVING!
MOSSES ALSO HAVE HARDY DENIZENS.
TARDIGRADES (OR "WATER BEARS")—WHICH ARE ROUGHLY THE SIZE OF A TYPED PERIOD—ARE BEASTS OF LEGEND FOR THEIR RESILIENCY.
IN HARSH CONDITIONS, THEY CURL INTO A *TUN*, AND METABOLIC FUNCTIONS COME TO A COMPLETE STOP.
IN THIS STATE, THEY CAN WITHSTAND THE VACUUM OF SPACE, PRESSURE SIX TIMES THAT OF THE DEEPEST PART OF THE OCEAN, LEVELS OF RADIATION 500 TIMES THE LETHAL HUMAN DOSE, AND VARIOUS TEMPERATURE EXTREMES.
WHEN FAVORABLE CONDITIONS RETURN, THEY UNCURL AND GO RIGHT BACK TO AMBLING AROUND THE MOSS AND SUCKING THE FLUIDS FROM PLANT CELLS.
HOWEVER, THIS EXCEPTIONAL DURABILITY IS NOT FOUND IN *ALL* SPECIES OF TARDIGRADES. WATER BEARS WHO LIVE IN LAKES OR OCEANS—STABLE ENVIRONMENTS—CANNOT SURVIVE SUCH EXTREMES. THESE DEATH-DEFYING POWERS ARE SPECIFIC TO SPECIES THAT LIVE IN MOIST FLORA, SUCH AS MOSSES.
SO WE CAN EXPECT WHOLE COMMUNITIES—NOT JUST THE MOSS—TO SURVIVE LONG, LONG COLD SPELLS, WHICH ARE ALSO "DRY"—WATER IS FROZEN!
EXTRAORDINARY!
TOK TOK
SHUFFLE SHUFFLE
BUFFALO JERKY

PLANTS THAT CAN CEASE METABOLIC FUNCTION AND RECOVER AGAIN ARE CALLED "RESURRECTION PLANTS."
MOSSES ARE CHAMPIONS IN THIS CATEGORY.
THE UNASSUMING GREEN FUZZ HAS POWERS THAT ARE TRULY MYTHIC IN SCOPE! LIKE CERTAIN DEITIES WHO POPULATE ANCIENT AND MODERN RELIGIONS.

AND FROGS CAN ALREADY DO WHAT COUNTLESS AUTHORS OF SCIENCE FICTION HAVE DREAMED OF DOING: TO BE CRYOGENICALLY FROZEN AND AWOKEN IN THE FUTURE, PERHAPS UPON ARRIVAL TO A DISTANT STAR.

WITH THEIR PHYSIOLOGIC SIMILARITIES TO HUMANS, FROGS CAN TEACH US A LOT ABOUT MEDICINE. ALREADY, RESEARCHERS HAVE IMPROVED THE TECHNOLOGY USED TO PRESERVE DONATED ORGANS AWAITING TRANSPLANT--BASED ON THE SLOW-FREEZING PROCESS OF FROGS.

CRYOGENIC FREEZING FOR PEOPLE IS STILL A DISTANT DREAM AT THIS POINT IN TIME. BUT WHO KNOWS WHAT THE FUTURE HOLDS?
I'M READY FOR MY SPACE MISSION.

I SAID CANCEL THE SOUTHWEST CARD AND KEEP THE UNITED ONE! LET ME TALK TO YOUR MANAGER!
IT APPEARS THAT WE LOST OUR AUDIENCE.
CHEW CHEW
THE PALEO DIET
BUFFALO JERKY
TASTE THE PRAIRIE
MAYBE BUD WAS LISTENING.
THE OLD PACMAN FROG IS RESILIENT. THIS WAS THE THIRD TIME I'VE MISTAKEN HIM FOR DEAD.
I'LL SAVE MY LITURGY TO RE-READ WHENEVER HE REALLY DIES. NOW, IT'S TIME TO PUT HIM BACK IN HIS TERRARIUM.
SHOULDN'T YOU USE A CONTAINER TO LIFT HIM?
HE'S STILL COMATOSE. I'LL JUST USE MY HANDS.
OUCH!
CHOMP!
RESURRECTED AND READY TO FIGHT, THAT'S OUR BUD.
THIS STORY IS DEDICATED TO YOU, MOM.
RWS
2016

THE TIRELESS TARDIGRADE!

by KEVIN KITE & MICHELLE McCAULEY

WHAT? YOU DON'T SEE IT? LOOK CLOSER...
THERE IT IS!
HI!
ALTHOUGH NEVER MORE THAN 1.5 mm LONG, THE TARDIGRADE HAS SUPER POWERS TO WITHSTAND FIERCE HEAT, FREEZING COLD, HIGH OXYGEN, OR NO OXYGEN!
PLUS, IT'S CUTE AND CUDDLY!
HEY! THAT'S NOT VERY DIGNIFYING!
PLOOP!

TELL MY STORY STRAIGHT!
OK
THE TARDIGRADE HAS BEEN AROUND AT LEAST 500 MILLION YEARS, SINCE BEFORE THE FIRST GREAT EXTINCTION- WAY BEFORE HUMANS ARRIVED!

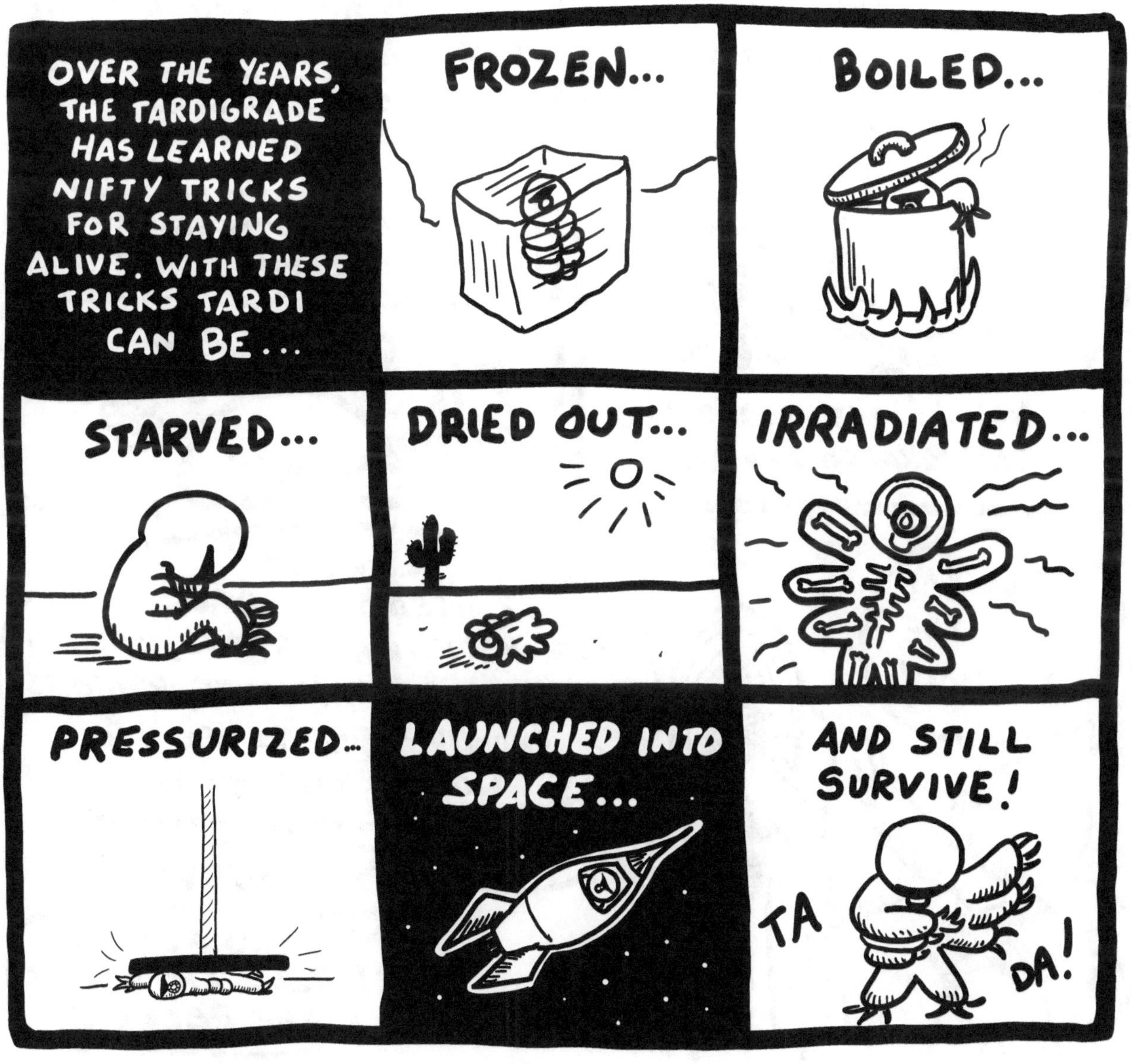
OVER THE YEARS, THE TARDIGRADE HAS LEARNED NIFTY TRICKS FOR STAYING ALIVE. WITH THESE TRICKS TARDI CAN BE...
FROZEN...
BOILED...
STARVED...
DRIED OUT...
IRRADIATED...
PRESSURIZED...
LAUNCHED INTO SPACE...
AND STILL SURVIVE!
TA
DA!

IN 1983, SCIENTISTS TOOK MOSS FROM ANTARCTICA, WRAPPED IT IN PLASTIC, ...
AND PUT IT IN DEEP FREEZE.
0 -10 -20 40 50 60
31 YEARS LATER, THEY PULLED THE MOSS OUT OF THE FREEZER AND ADDED WATER.
AND VOILA! THEY FOUND TWO LIVE TARDIGRADES AND AN EGG.
THE EGG HATCHED, AND THESE THREE TARDIGRADES WENT ON TO REPRODUCE OTHERS
FAMILY PORTRAIT 2014
TELL 'EM ABOUT THE RADIATION!
SCIENTISTS ZAPPED TARDIGRADES WITH RADIATION — X-RAYS, GAMMA RAYS, ALPHA PARTICLES, PROTONS, AND UV — JUST TO SEE IF IT WOULD KILL THEM.
AND...
YOU SURVIVED.
YOU BETCHA! TELL 'EM ABOUT THE ROCKETS!

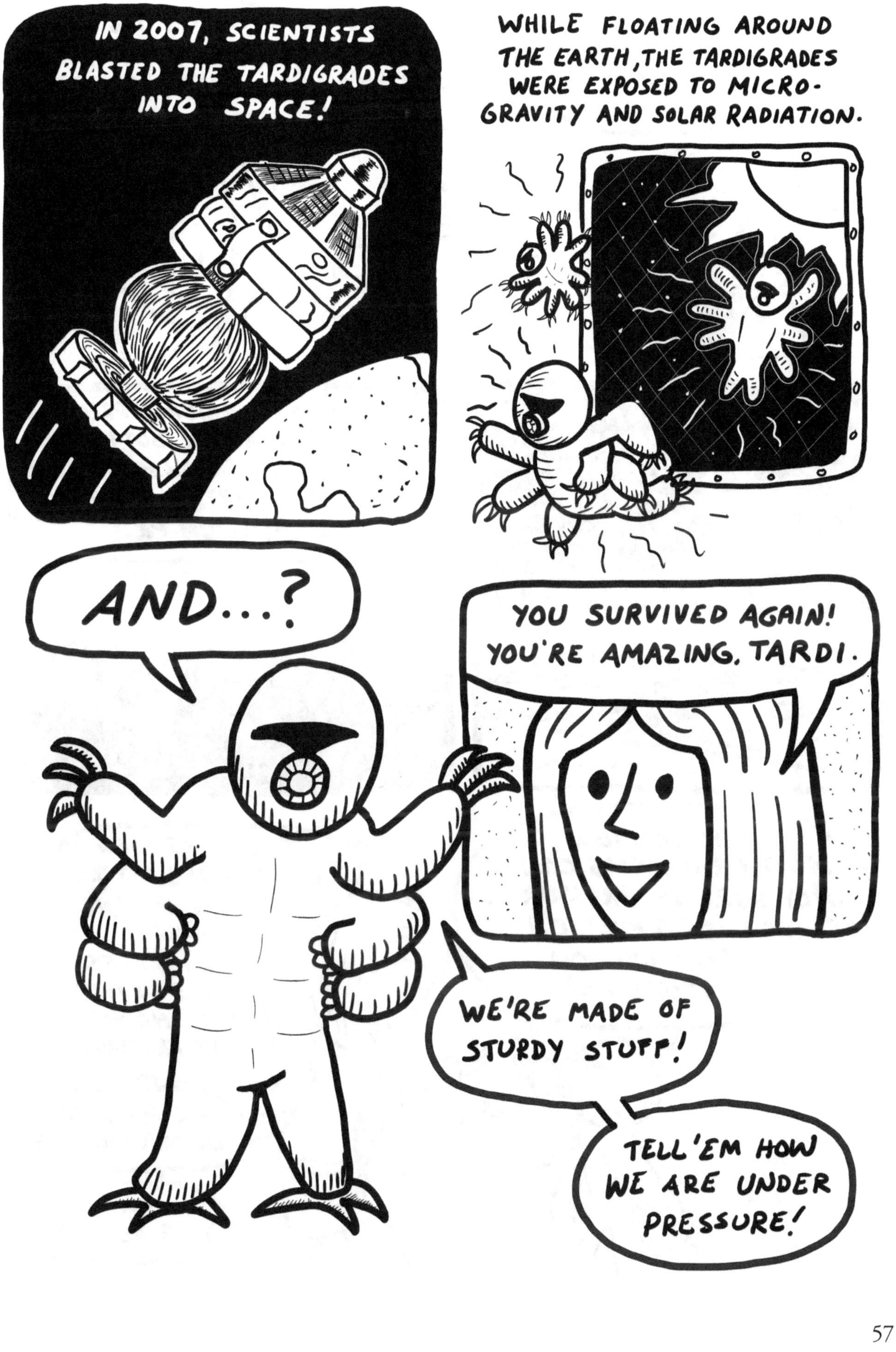
IN 2007, SCIENTISTS BLASTED THE TARDIGRADES INTO SPACE!
WHILE FLOATING AROUND THE EARTH, THE TARDIGRADES WERE EXPOSED TO MICRO-GRAVITY AND SOLAR RADIATION.
AND...?
YOU SURVIVED AGAIN! YOU'RE AMAZING, TARDI.
WE'RE MADE OF STURDY STUFF!
TELL'EM HOW WE ARE UNDER PRESSURE!

I SAW YOU IN A HIGH STAKES CARD GAME, AND YOU DIDN'T BREAK A SWEAT
I MEAN REAL PRESSURE.
OH! HOW YOU CAN LIVE WHERE HUMANS CAN'T.
PRESSURE IS MEASURED IN PASCALS, ABBREVIATED Pa. ONE Pa IS THE PRESSURE OF ONE DOLLAR BILL, LYING ON A TABLE.
ONE DOLLAR
THE HIGHEST NATURAL PRESSURE ON EARTH IS AT THE BOTTOM OF THE MARIANA TRENCH, 7 MILES BELOW THE EARTH'S SURFACE-110 MILLION Pa.
36,200 FT
STUDIES SHOW MOST ORGANISMS CAN ONLY SURVIVE PRESSURES OF ABOUT 600 MILLION Pa. SO WHAT DID YOU SCIENTISTS DO?
THEY EXPOSED TARDIGRADES TO 7.5 BILLION Pa AND FOUND TARDIGRADES COULD SURVIVE FOR 12 HOURS AT THAT PRESSURE.
WE ARE COOL UNDER PRESSURE
WHAT'S YOUR SECRET?

HOLD ON! I WANT A COAT FOR THIS.
IF IT GETS TOO DRY, OR TOO SALTY, OR IF THE OXYGEN LEVEL FALLS TOO LOW, OR FOOD DISAPPEARS—POOF! WE GO INTO CRYPTOBIOSIS!
CRYPTO-WHAT?
IT MEANS "HIDDEN LIFE." ONE WAY I DO IT IS THROUGH ANHYDROBIOSIS—BASICALLY, I DRY OUT! I DUMP ALL MY WATER, SLOW MY METABOLISM TO 0.01% OF NORMAL, AND CURL UP IN A BALL CALLED A TUN
IN OUR TUN STATE, TARDIGRADES CAN SURVIVE ALMOST ANYTHING!
HERE'S A COOL THING. ANYONE CAN WAKE A TUN WITH WATER.

SPLASH!
HEY! WHAT'S THE DEAL, DUDE?
WE'RE NOT DONE HERE.
I THINK THEY GET THE POINT. I CAN SURVIVE ANYTHING, AS LONG AS YOU LET ME SLEEP.
YOU DON'T HAVE TO SLEEP AS A TUN TO SURVIVE. YOU CAN ALSO USE CRYOBIOSIS— SOMEHOW, YOU MANAGE TO FREEZE WITHOUT GETTING HURT.
WE'RE NOT SURE HOW YOU DO IT.
HEY! YOU'RE NOT TALKIN' MORE TESTS, ARE YOU?
THERE'S A FEW MORE THINGS THEY OUGHT TO KNOW.

FIRST, TARDIGRADES LIVE EVERYWHERE...
IN THE SEA.
ON THE LAND.
IN FRESHWATER.
BUT TARDIGRADES WHO LIVE IN WATER AREN'T AS GOOD AT SURVIVING EXTREME CONDITIONS AS THEIR TERRESTRIAL COUSINS – THEY ARE THE ONES WITH THE CRYOBIOTIC TRICKS.
YEAH. WE'RE SPECIAL THAT WAY.
SO, TARDIGRADES COULD SURVIVE AT THE TOP OF MOUNT EVEREST OR THE BOTTOM OF THE MARINA TRENCH.

BUT WE'RE MOST LIKELY TO BE IN YOUR BACKYARD! WE LOVE MOSS AND LICHEN.
SO IF YOU HAVE A MOSSY TREE, YOU CAN LOOK FOR TARDI.
JUST GO OUT AND COLLECT SOME MOSS FROM THE NEARBY TREE.
PUT IT IN A BOWL, COVER IT WITH WATER AND WAIT 24 HOURS.
THEN, LOOK FOR LITTLE MOVING BEINGS IN THE WATER.
PUT THE MOVING THINGS ON A SLIDE UNDER A MICROSCOPE.
40x

YOU'LL KNOW THE TARDIGRADE BY ITS CUTE CURLY CLAWS, AND IT'S EIGHT WIGGLY LEGS. IT WON'T BE MOVING QUICKLY BECAUSE "TARDIGRADE" MEANS "SLOW-WALKING."
MAYBE THEY'LL BE MY COUSINS! SAY HELLO FOR ME.
HEY TARDI — WHERE'RE YOU GOING?
I GOTTA SEE A DOCTOR.
AND THAT IS THE AMAZING STORY OF THE TIRELESS. TARDIGRADE.
POLICE BOX
VWROOP! VWROOP! VWROOP!
FIN

SPRATTY 2016

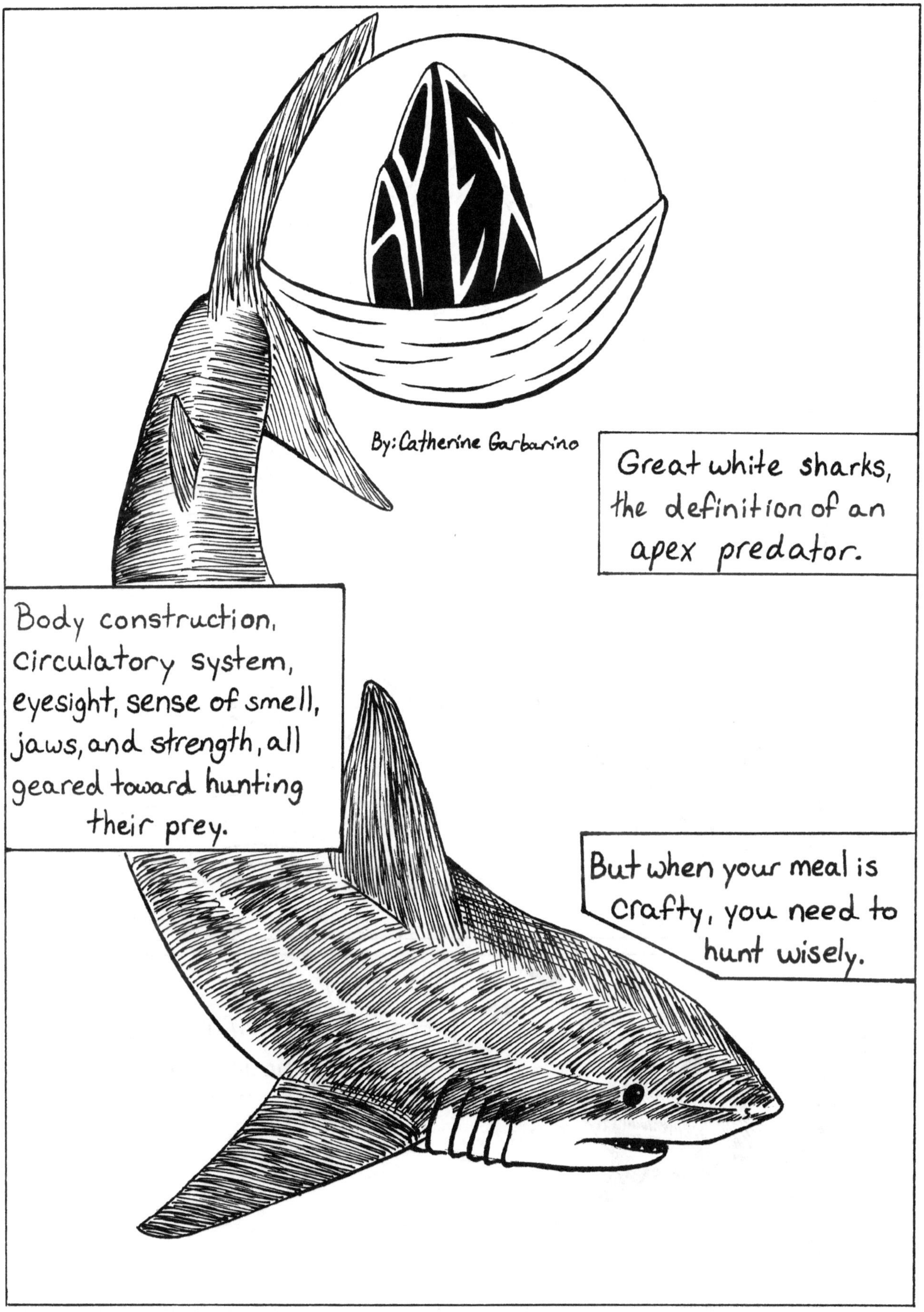
APEX
By: Catherine Garbarino
Great white sharks, the definition of an apex predator.
Body construction, circulatory system, eyesight, sense of smell, jaws, and strength, all geared toward hunting their prey.
But when your meal is crafty, you need to hunt wisely.

California: Farallon Islands

Seals arrive in the Fall

Hungry sharks swim in from the Pacific Ocean where they have summered.

The shark attacks the seal from behind and below
The shark then swims down to the bottom to drown the seal
If the blood loss and massive trauma from the bite don't do it first that is.

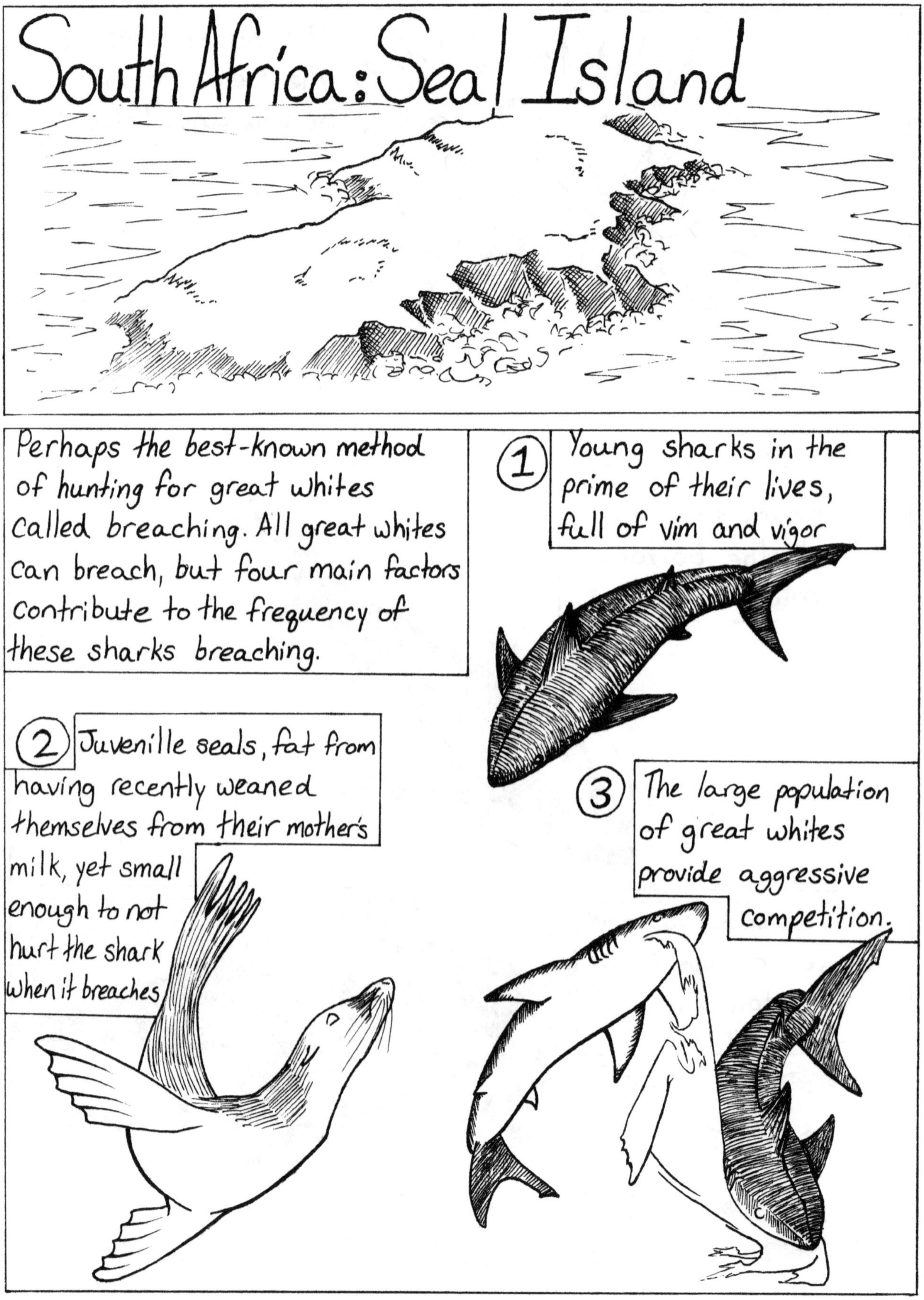
South Africa: Seal Island
Perhaps the best-known method of hunting for great whites called breaching. All great whites can breach, but four main factors contribute to the frequency of these sharks breaching.
1
Young sharks in the prime of their lives, full of vim and vigor
2
Juvenille seals, fat from having recently weaned themselves from their mother's milk, yet small enough to not hurt the shark when it breaches
3
The large population of great whites provide aggressive competition.

4
A good layout for breaching
W
N
S
E
The West side is deeper than the East
The South side is deeper than the North
This creates a primary departure/return path for seals, and a bountiful hunting ground for the sharks.

Mexico: Guadalupe Island
The waters surrounding this island are some of the clearest you can find
With clarity up to 150ft, seals can swim with confidence in the knowlege that they can spot a predator of any size coming and act accordingly.
But great whites have solved this problem

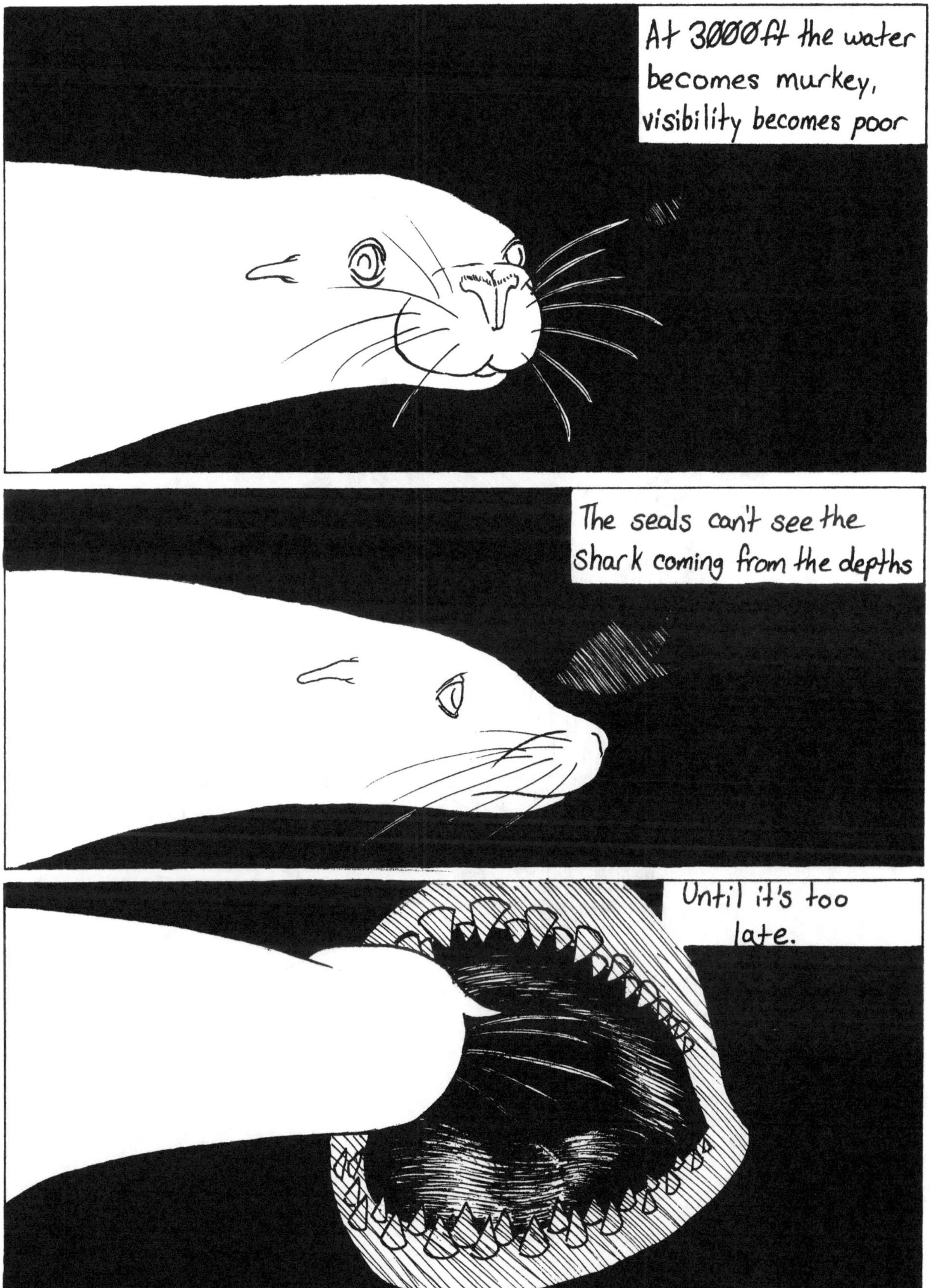
At 3000ft the water becomes murkey, visibility becomes poor
The seals can't see the shark coming from the depths
Until it's too late.

Great white sharks:
Apex predator.
Every aspect of its anatomy geared toward survival.
Toward hunting. And when your meal is crafty
You need to be craftier.

INTRODUCING
SALMON
Kit Fraser
A DETERMINED FISH
THAT NEEDS OUR HELP.

A. K. A.
Salmonidae
OR AS YOU MIGHT CALL IT:
DINNER.
IN THE PAST TWENTY YEARS, SALMON CONSUMPTION HAS DOUBLED, MAKING IT THE THIRD MOST POPULAR SEAFOOD IN THE UNITED STATES.
THIS RISE IN POPULARITY IS DUE TO SALMON'S HEALTHY OMEGA-3 FAT CONTENT, DELICIOUS TASTE, AND INCREASED AVAILABILITY BECAUSE OF SALMON FARMING.
(MORE ON THAT SOON...)
But FIRST things FIRST...
WHAT MAKES SALMON SO SPECIAL AND WORTHY OF OUR ATTENTION?

EACH SALMON'S LIFE BEGINS AND ENDS IN THE SAME PLACE: ONE SPECIFIC AREA OF A COOL, CLEAN RIVER WITH PLENTY OF TIMBER COVER AND OXYGEN-RICH WATER.

Life Cycle
EGGS
THOUSANDS OF EGGS* ARE DEPOSITED AND FERTILIZED UP RIVER IN SHALLOW, CALMER AREAS.
*80% OF EGGS DIE.
ALEVIN
AFTER 4-6 WEEKS. ALEVIN HATCH, STILL FEEDING FROM THEIR YOLK SAC.
FRY
ONCE THE YOLK SAC IS ABSORBED, ALEVIN BECOME FRY. THEY EAT MICROSCOPIC STREAM LIFE AND CAN MANEUVER THROUGH THE WATER.
PARR
AFTER 1 YEAR, FRY BECOME PARR. THEY SPEND 1-5 YEARS IN THE RIVER, GROWING STRONGER ON A DIET OF AQUATIC INSECTS.
SMOLT
IN LATE SPRING PARR SALMON TRANSFORM. THEIR MARKINGS CHANGE AND THEY BEGIN, FOR THE FIRST TIME, TO SWIM DOWNSTREAM TO THE OCEAN. TO PREPARE, SMOLT LEARN HOW TO REGULATE SALT AND WATER IN THEIR BODIES.
ADULT
FOR 1-4 YEARS THEY TRAVEL HUGE DISTANCES TO FEED ON SMALL FISH. SALMON RETURN TO THEIR NATIVE STREAMS TO SPAWN AND DIE.
IN THE END, ABOUT 2% OF HATCHED SALMON REACH ADULTHOOD!

SALMON REALLY PROVE THEMSELVES AS AWESOME CREATURES AS THEY REACH ADULTHOOD.
THEY TRAVEL HUNDREDS OF MILES TO FERTILE FEEDING GROUNDS LIKE ICELAND AND RUSSIA.
Sockeye
AND WHEN IT'S TIME*, THEY HEAD HOME TO FULFILL THEIR MISSION OF SPAWNING IN THEIR OWN BIRTHPLACE. THE JOURNEY IS HARD— MANY MILES OF SWIMMING AGAINST THE CURRENT. SALMON STOP EATING ONCE THEY LEAVE THE OCEAN, AND THEY FACE MANY PREDATORS.
*LARGELY DETERMINED BY WHICH SALMON SPECIES THEY ARE.

These predators
are among those that need salmon.
EXISTENCE IN PLACES LIKE ALASKA RELIES ALMOST EXCLUSIVELY ON THE WILD SALMON RUNS.
TREE GROWTH IS 3X HIGHER IN AREAS WITH SALMON SPAWNING.
WHOLE ECOSYSTEMS (NOT TO MENTION CULTURES THOUSANDS OF YEARS OLD!) HINGE ON SALMON COMPLETING THEIR MISSION.
And yet...

MODERN SOCIETY

IS AT ODDS WITH SALMON'S SUCCESS AND ALL THE PEOPLE, PLACES, AND ANIMALS THAT RELY ON THEM.

*WATERS WARMING FROM CLIMATE CHANGE NEGATIVELY IMPACT SALMON.

*MINING PROJECTS LIKE PEBBLE MINE IN ALASKA THREATEN TO WIPE OUT THE FEW REMAINING WILD SALMON RUNS.

*DAMNS PREVENT SALMON FROM REACHING THEIR SPAWNING GROUNDS AND PLAY A HUGE ROLE IN THE DECLINE OF WILD ATLANTIC SALMON.

WHAT CAN WE DO?

*TAKE STEPS TO SLOW GLOBAL WARMING AND WATER POLLUTION, LIKE AVOID USING PLASTIC AND DRIVING WHEN YOU CAN WALK OR RIDE A BIKE.

*WRITE YOUR REPRESENTATIVES TO EXPRESS YOUR CONCERNS ABOUT DRILLING AND MINING THAT ENDANGER SALMON. TELL THEM YOU SUPPORT THE REMOVAL OF INACTIVE DAMNS AND THE RESTORATION OF RIVERS.

WILD ATLANTIC SALMON HAVE NEARLY BEEN FISHED OUT OF EXISTENCE:
*BY 1999, PACIFIC SALMON WERE EXTINCT IN 40% OF RIVERS WHERE THEY USED TO THRIVE.
*ALASKAN SALMON OUTNUMBER ALASKANS 1,500:1 BUT GLOBALLY THE RATIO OF SALMON TO HUMANS IS 1:7. THERE'S SIMPLY NOT ENOUGH TO GO AROUND!
Overfishing
WHAT CAN WE DO?
*NEVER EAT WILD ATLANTIC SALMON. THEY'RE VERY ENDANGERED.
*VISIT SEAFOODWATCH.ORG FOR SUSTAINABLE, HEALTHY ALTERNATIVES.
Salmon Farming
FARMED SALMON IS OFTEN THOUGHT OF AS A CHEAP, GOOD ALTERNATIVE TO ITS WILD COUSIN BUT THEY'RE PROBLEMATIC:
*ESCAPED FARMED SALMON CONSUME WILD SALMON'S RESOURCES.
*THEY CAN SPREAD SICKNESS AND REDUCE GENETIC DIVERSITY.
*TO GROW 1 LB OF SALMON, YOU NEED 4 LBS OF SMALLER FISH. THIS UNSUSTAINABLE RATIO HAS HURT THE POPULATION OF SMALL WILD FISH.
The BIGGEST PROBLEM SALMON FACE IS apathy...
DO WE CARE ENOUGH TO SAVE SALMON & EVERYTHING THAT DEPENDS ON THEM?
WHAT CAN WE DO?
*AVOID EATING FARMED SALMON, OPTING FOR WILD, SUSTAINABLY HARVESTED SALMON INSTEAD.
*IN GENERAL TRY TO REDUCE THE AMOUNT OF SALMON YOU EAT.
*LOOK FOR THE MARINE STEWARDSHIP COUNCIL'S BLUE LABEL BEFORE BUYING FISH.
*TALK TO YOUR FRIENDS AND FAMILY. SPREAD THE WORD!

SALMON'S PROBLEMS CAN FEEL DISTANT.
BAR
Especially FOR THOSE OF US FAR AWAY, WHERE WILD SALMON RUNS WERE DESTROYED GENERATIONS AGO.
BUT OUR ACTIONS STILL MATTER.
WE DON'T NEED TO MAKE THE SAME MISTAKES.
GIVEN HALF A CHANCE, WILD SALMON WILL THRIVE AGAIN.
Let's give them that chance.

SO LET'S GET TO THE BOTTOM (THE TOP?) OF THIS, SHALL WE?

FACT #1: THEY ARE TALL.
THE AVERAGE HEIGHT OF AN ADULT GIRAFFE IS BETWEEN 14 - 19 FEET.
THEIR LEGS MEASURE UP TO 6 FEET HIGH ALONE!
CLK CLK
SO THEY COULD EASILY PEER INTO A SECOND STORY WINDOW!
THERE'S A GIRAFFE AT THE WINDOW.
YOU'LL SAY ANYTHING TO AVOID READING MY BLOG!

FACT #2: THEY HAVE LONG NECKS.
A GIRAFFE'S NECK WEIGHS ***ABOUT 600 POUNDS*** AND IS ***ABOUT 6 FEET LONG***!
ALL THIS, EVEN THOUGH THEY HAVE SEVEN CERVICAL VERTEBRAE -- ***THE SAME NUMBER*** FOUND IN ***A HUMAN NECK***!
A GIRAFFE'S VERTEBRAE CAN EACH MEASURE ***10 INCHES*** IN LENGTH.
GIRAFFES USE THEIR LONG, POWERFUL NECKS TO ENGAGE IN A PRACTICE CALLED "NECKING."
NOT THIS. CUTE, THOUGH.

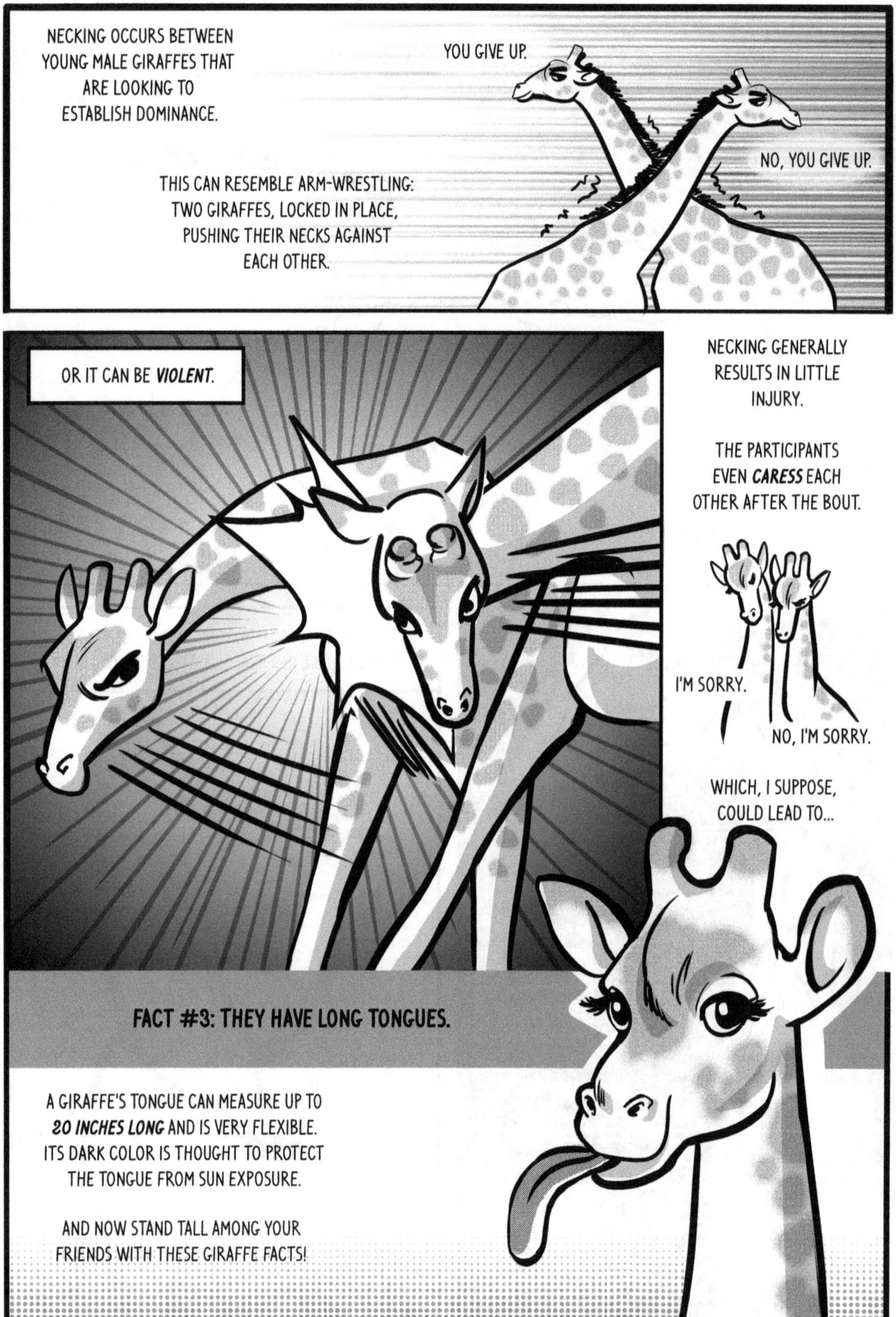
NECKING OCCURS BETWEEN YOUNG MALE GIRAFFES THAT ARE LOOKING TO ESTABLISH DOMINANCE.
YOU GIVE UP.
NO, YOU GIVE UP.
THIS CAN RESEMBLE ARM-WRESTLING: TWO GIRAFFES, LOCKED IN PLACE, PUSHING THEIR NECKS AGAINST EACH OTHER.
OR IT CAN BE ***VIOLENT***.
NECKING GENERALLY RESULTS IN LITTLE INJURY.
THE PARTICIPANTS EVEN ***CARESS*** EACH OTHER AFTER THE BOUT.
I'M SORRY.
NO, I'M SORRY.
WHICH, I SUPPOSE, COULD LEAD TO...
FACT #3: THEY HAVE LONG TONGUES.
A GIRAFFE'S TONGUE CAN MEASURE UP TO ***20 INCHES LONG*** AND IS VERY FLEXIBLE. ITS DARK COLOR IS THOUGHT TO PROTECT THE TONGUE FROM SUN EXPOSURE.
AND NOW STAND TALL AMONG YOUR FRIENDS WITH THESE GIRAFFE FACTS!

WHILE OUTER SPACE HAS ITS FASCINATIONS SO DOES THE OCEAN HERE ON EARTH.
THE DEEP, COLD, HEAVY-PRESSURED ABYSS OF THE FOREVER UTTERLY DARK OCEAN BED...
... IS INHABITED BY THE WEIRDEST AND MEANEST CREATURES ON THE PLANET.
BUT VERY LITTLE IS KNOWN ABOUT THESE CREATURES.
THE DEEPEST POINT IN THE OCEAN IS 7 MILES DOWN OR 36,000 ft (40m), AND TO-DATE ONLY A FEW PEOPLE HAVE BEEN DOWN THERE.
NOW THIS IS A SMALL LIST OF SUCH CREATURES, SOME MADE IT FOR THEIR FEROCITY WHILE SOME JUST FOR THEIR LOOKS...
YES, YES I'M INTO LOOKS #dealwithit!
SO, LET'S DIVE IN AND FIND SOME...
TERRORS OF THE DEEP SEA
BY Shashwat Mishra

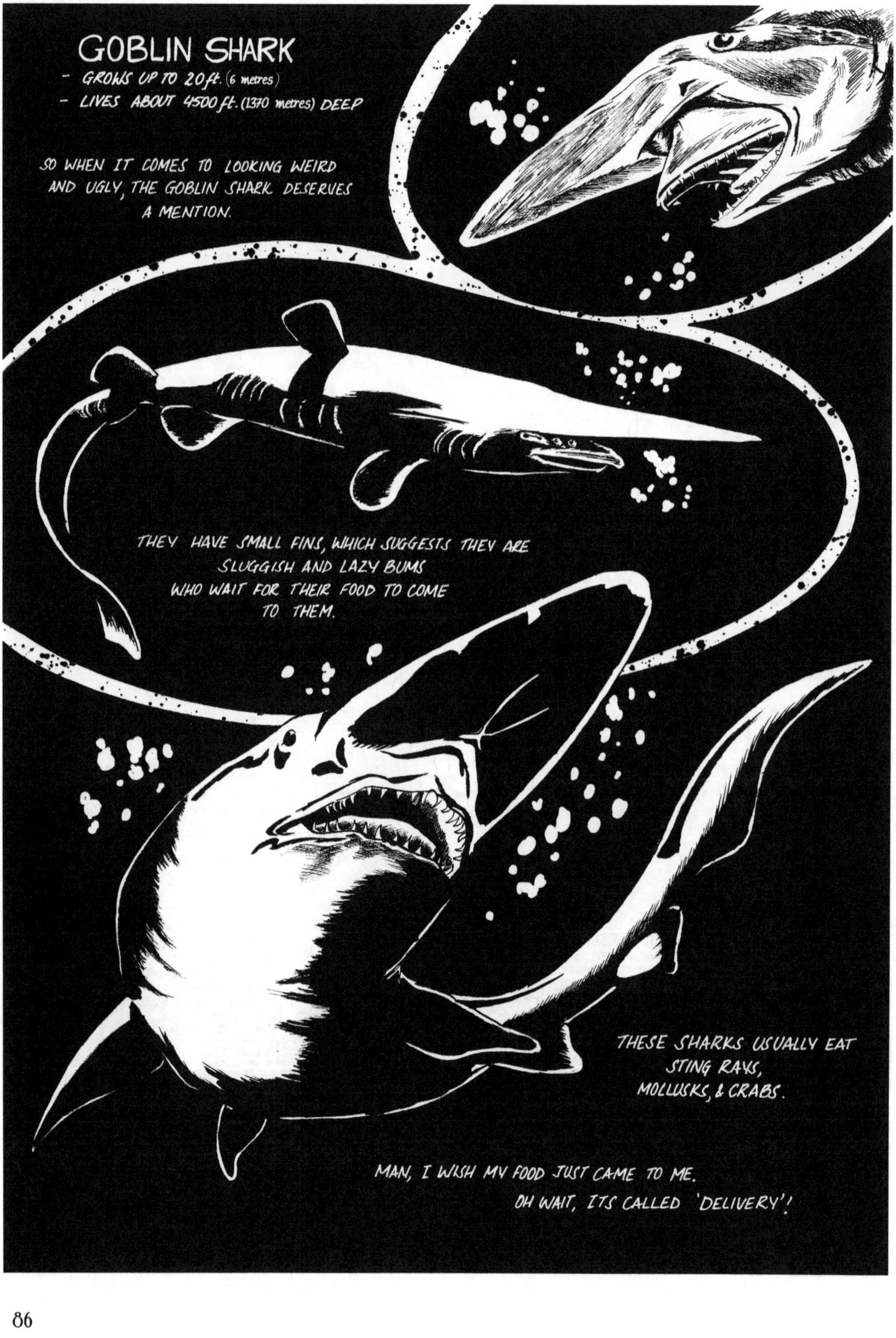
GOBLIN SHARK
- GROWS UP TO 20 ft. (6 metres)
- LIVES ABOUT 4500 ft. (1370 metres) DEEP
SO WHEN IT COMES TO LOOKING WEIRD AND UGLY, THE GOBLIN SHARK DESERVES A MENTION.
THEY HAVE SMALL FINS, WHICH SUGGESTS THEY ARE SLUGGISH AND LAZY BUMS WHO WAIT FOR THEIR FOOD TO COME TO THEM.
THESE SHARKS USUALLY EAT STING RAYS, MOLLUSKS, & CRABS.
MAN, I WISH MY FOOD JUST CAME TO ME. OH WAIT, ITS CALLED 'DELIVERY'!

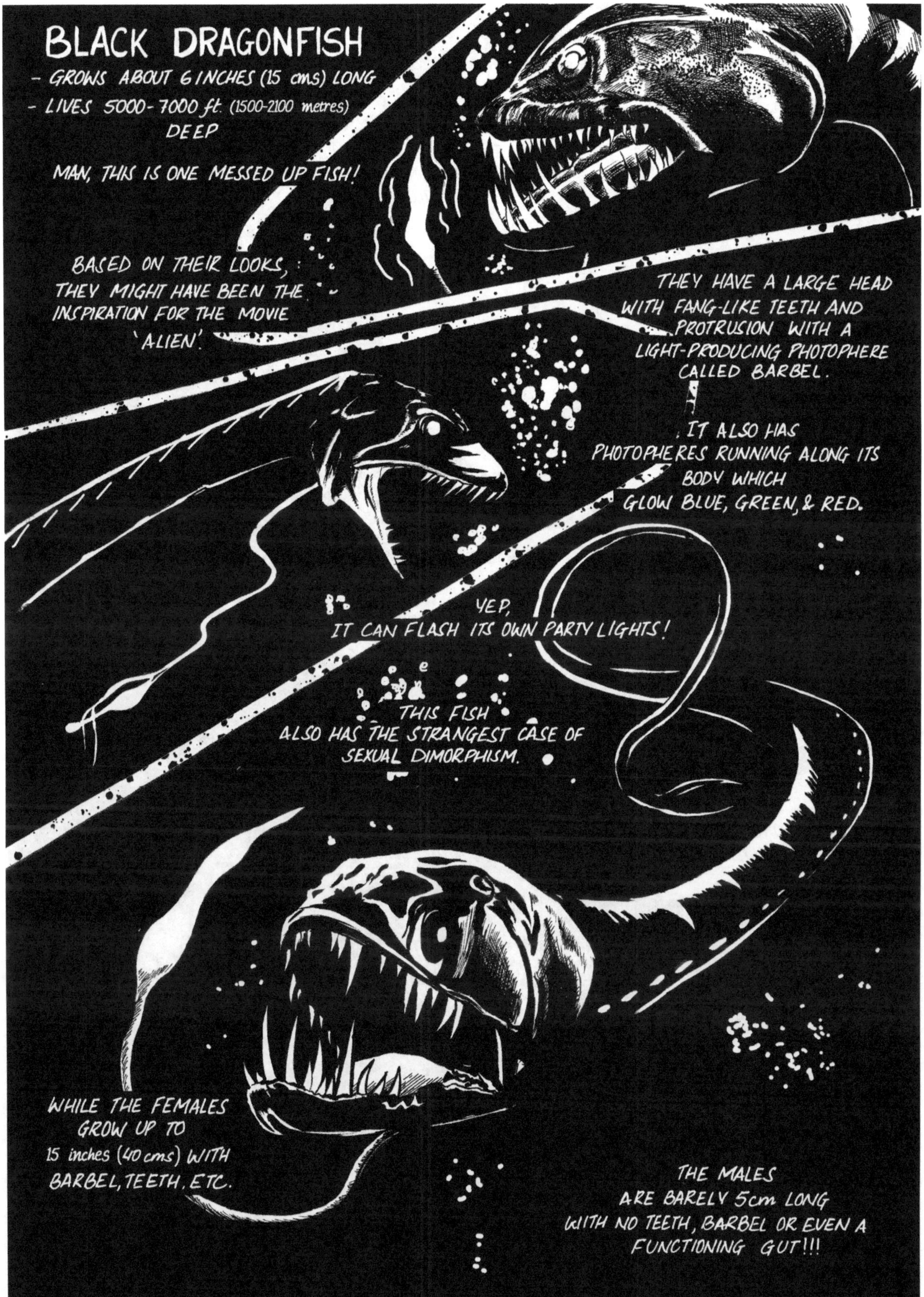
BLACK DRAGONFISH
- GROWS ABOUT 6 INCHES (15 cms) LONG
- LIVES 5000-7000 ft. (1500-2100 metres) DEEP
MAN, THIS IS ONE MESSED UP FISH!
BASED ON THEIR LOOKS, THEY MIGHT HAVE BEEN THE INSPIRATION FOR THE MOVIE 'ALIEN'.
THEY HAVE A LARGE HEAD WITH FANG-LIKE TEETH AND PROTRUSION WITH A LIGHT-PRODUCING PHOTOPHERE CALLED BARBEL.
IT ALSO HAS PHOTOPHERES RUNNING ALONG ITS BODY WHICH GLOW BLUE, GREEN, & RED.
YEP, IT CAN FLASH ITS OWN PARTY LIGHTS!
THIS FISH ALSO HAS THE STRANGEST CASE OF SEXUAL DIMORPHISM.
WHILE THE FEMALES GROW UP TO 15 inches (40 cms) WITH BARBEL, TEETH, ETC.
THE MALES ARE BARELY 5cm LONG WITH NO TEETH, BARBEL OR EVEN A FUNCTIONING GUT!!!

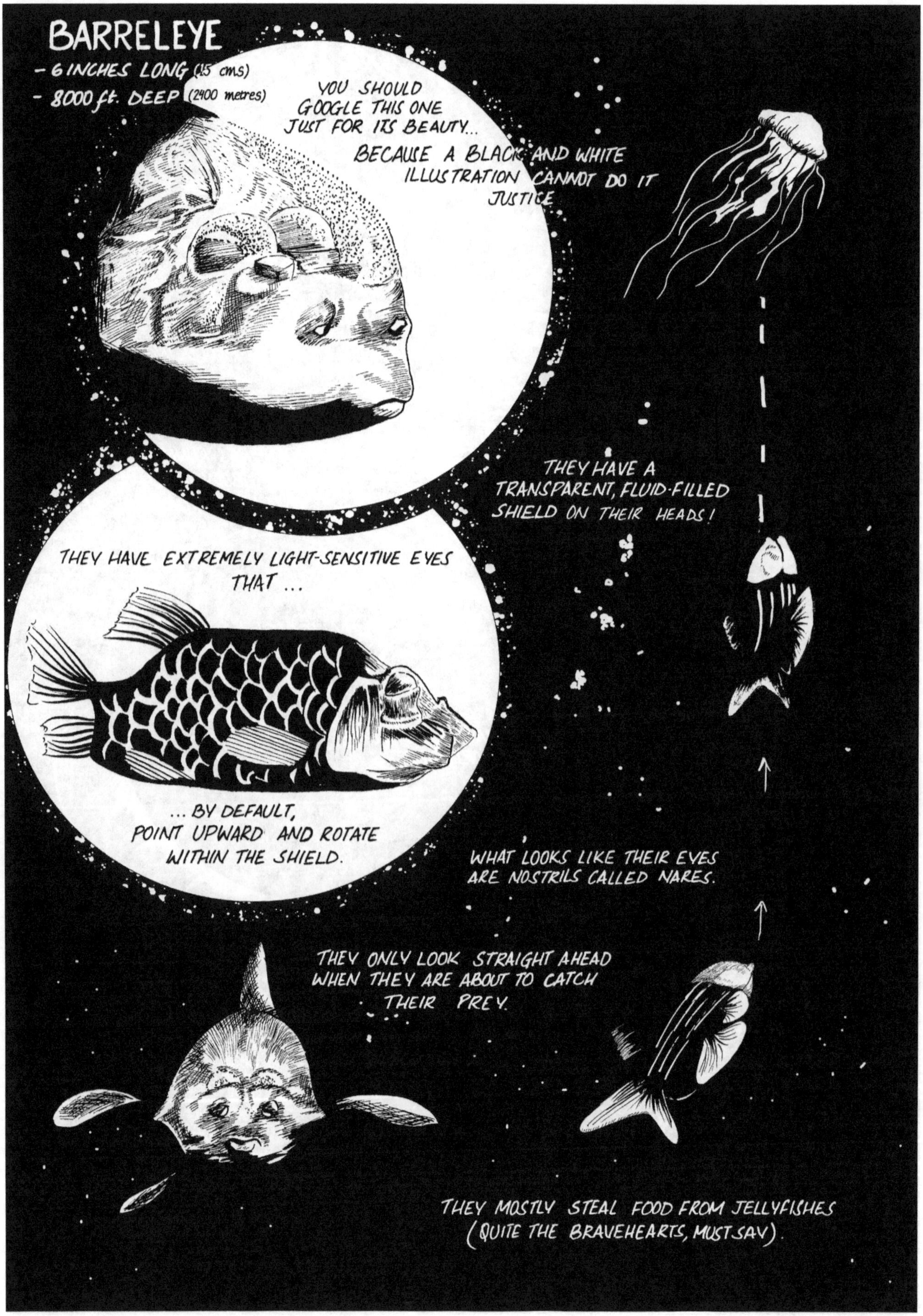
BARRELEYE
- 6 INCHES LONG (15 cms)
- 8000 ft. DEEP (2400 metres)
YOU SHOULD GOOGLE THIS ONE JUST FOR ITS BEAUTY...
BECAUSE A BLACK AND WHITE ILLUSTRATION CANNOT DO IT JUSTICE
THEY HAVE A TRANSPARENT, FLUID-FILLED SHIELD ON THEIR HEADS!
THEY HAVE EXTREMELY LIGHT-SENSITIVE EYES THAT ...
... BY DEFAULT, POINT UPWARD AND ROTATE WITHIN THE SHIELD.
WHAT LOOKS LIKE THEIR EYES ARE NOSTRILS CALLED NARES.
THEY ONLY LOOK STRAIGHT AHEAD WHEN THEY ARE ABOUT TO CATCH THEIR PREY.
THEY MOSTLY STEAL FOOD FROM JELLYFISHES (QUITE THE BRAVEHEARTS, MUST SAY).

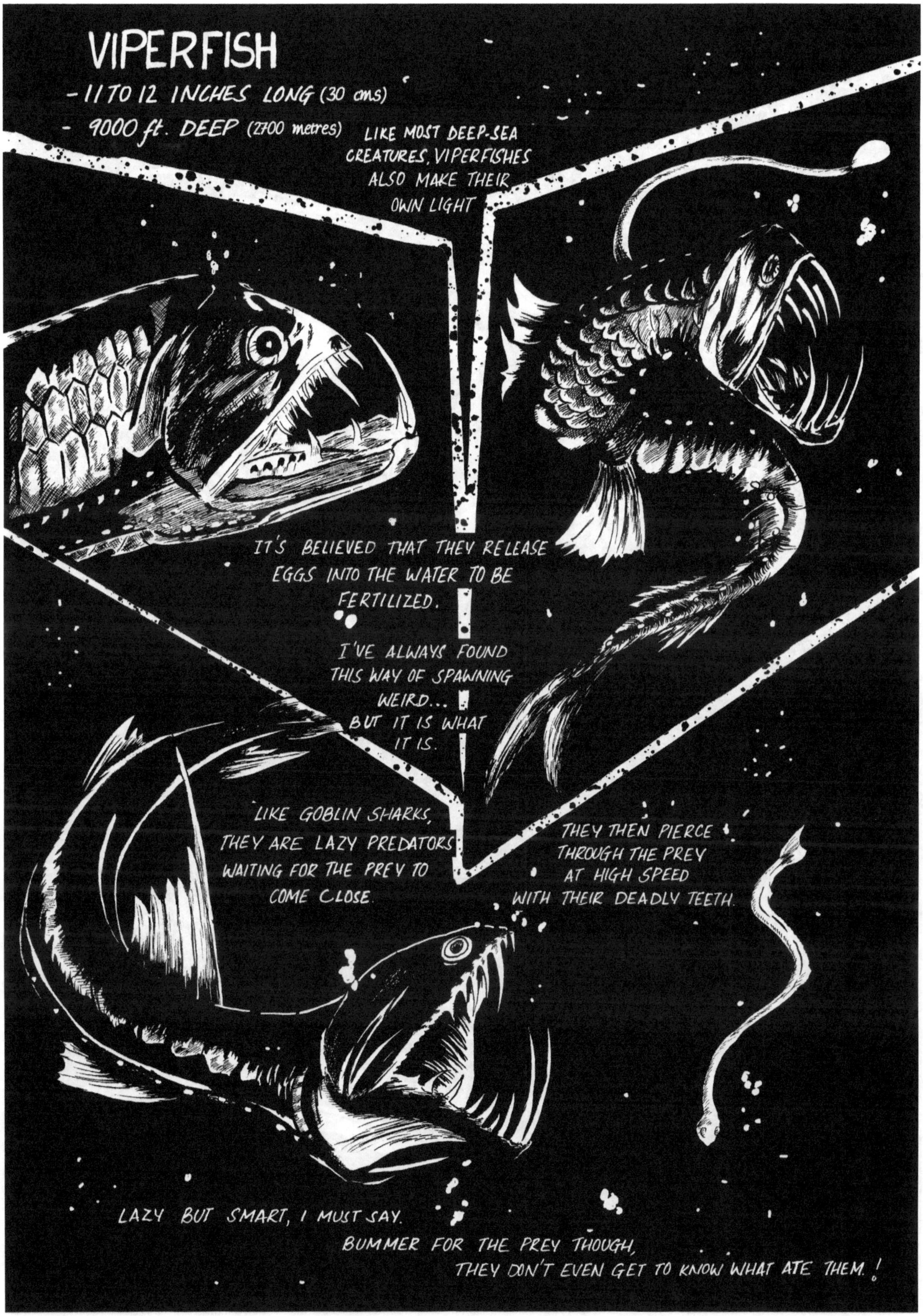
VIPERFISH
- 11 TO 12 INCHES LONG (30 cms)
- 9000 ft. DEEP (2700 metres)
LIKE MOST DEEP-SEA CREATURES, VIPERFISHES ALSO MAKE THEIR OWN LIGHT
IT'S BELIEVED THAT THEY RELEASE EGGS INTO THE WATER TO BE FERTILIZED.
I'VE ALWAYS FOUND THIS WAY OF SPAWNING WEIRD... BUT IT IS WHAT IT IS.
LIKE GOBLIN SHARKS, THEY ARE LAZY PREDATORS WAITING FOR THE PREY TO COME CLOSE.
THEY THEN PIERCE THROUGH THE PREY AT HIGH SPEED WITH THEIR DEADLY TEETH.
LAZY BUT SMART, I MUST SAY.
BUMMER FOR THE PREY THOUGH,
THEY DON'T EVEN GET TO KNOW WHAT ATE THEM!

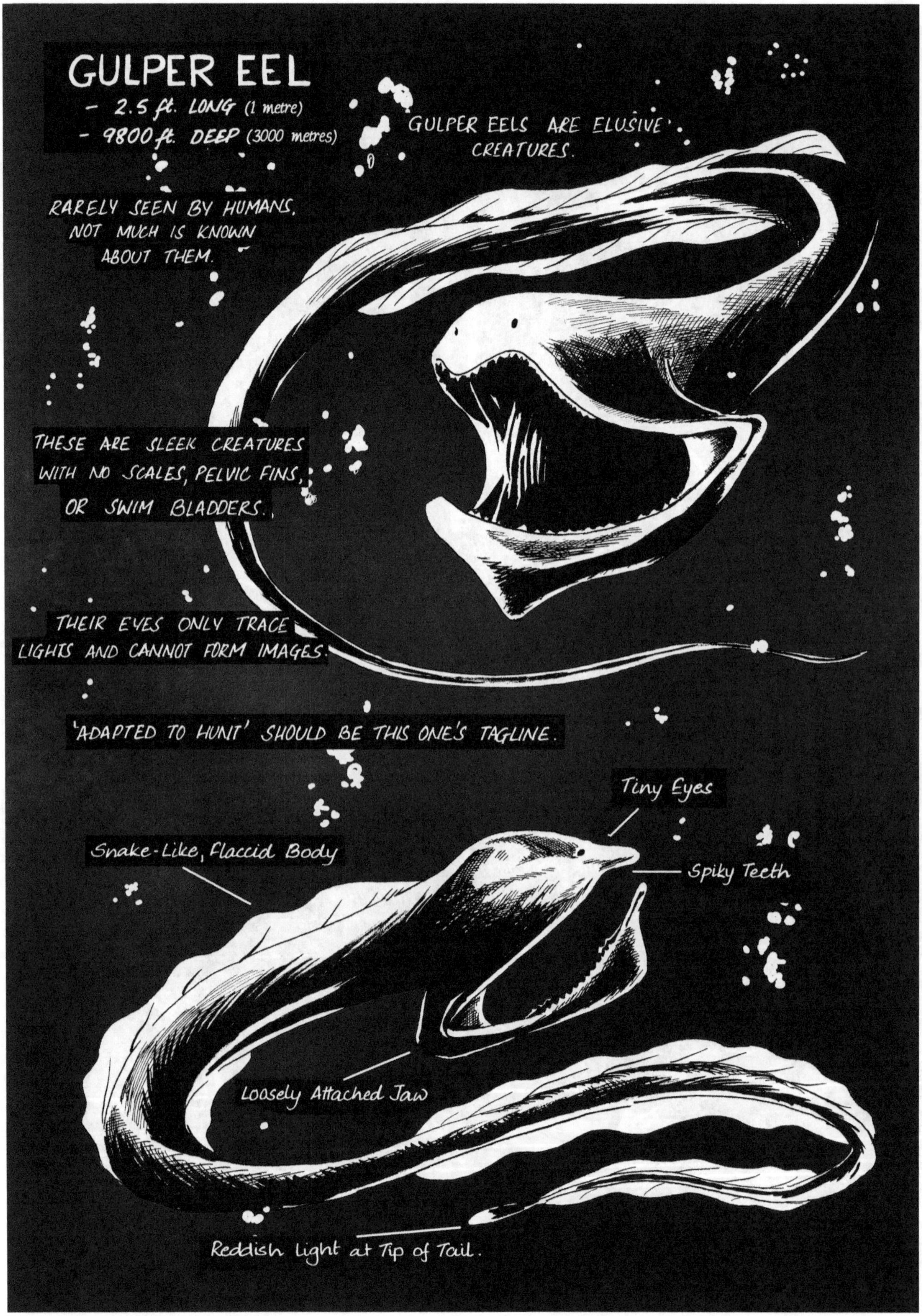
GULPER EEL
- 2.5 ft. LONG (1 metre)
- 9800 ft. DEEP (3000 metres)
GULPER EELS ARE ELUSIVE CREATURES.
RARELY SEEN BY HUMANS. NOT MUCH IS KNOWN ABOUT THEM.
THESE ARE SLEEK CREATURES WITH NO SCALES, PELVIC FINS, OR SWIM BLADDERS.
THEIR EYES ONLY TRACE LIGHTS AND CANNOT FORM IMAGES.
'ADAPTED TO HUNT' SHOULD BE THIS ONE'S TAGLINE.
Tiny Eyes
Snake-Like, Flaccid Body
Spiky Teeth
Loosely Attached Jaw
Reddish Light at Tip of Tail.

SWALLOWER
- 10 INCHES LONG (25 cms)
- 9000 ft. DEEP (2700 metres)
THIS FISH HAS A MASSIVE, STRETCHY STOMACH.
IT CAN SWALLOW PREY TWICE ITS SIZE AND 10 TIMES ITS WEIGHT.
WISH I COULD DO THAT WITH ICE CREAM. AND BELIEVE ME, I'VE TRIED.
BUT SWALLOWING TOO BIG OF PREY MIGHT KILL IT FROM INDIGESTION.
AN OVERSIZED MEAL MAY LIE IN ITS STOMACH DECOMPOSING, RELEASING GASES THAT BLOAT THE STOMACH, PUSHING IT TO THE WATER'S SURFACE.
ROUGH!

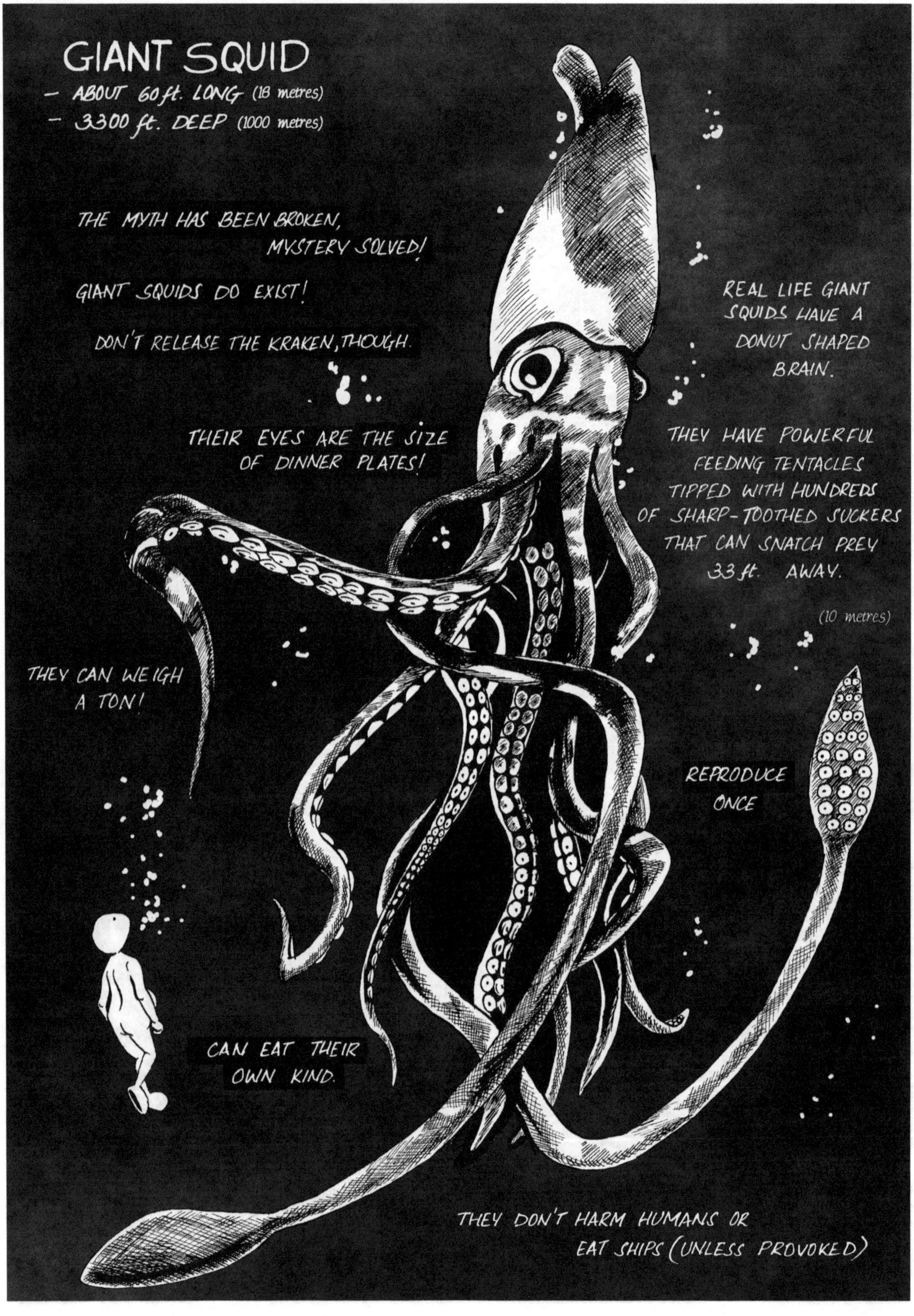
GIANT SQUID
– ABOUT 60 ft. LONG (18 metres)
– 3300 ft. DEEP (1000 metres)
THE MYTH HAS BEEN BROKEN, MYSTERY SOLVED!
GIANT SQUIDS DO EXIST!
DON'T RELEASE THE KRAKEN, THOUGH.
REAL LIFE GIANT SQUIDS HAVE A DONUT SHAPED BRAIN.
THEIR EYES ARE THE SIZE OF DINNER PLATES!
THEY HAVE POWERFUL FEEDING TENTACLES TIPPED WITH HUNDREDS OF SHARP-TOOTHED SUCKERS THAT CAN SNATCH PREY 33 ft. AWAY.
(10 metres)
THEY CAN WEIGH A TON!
REPRODUCE ONCE
CAN EAT THEIR OWN KIND.
THEY DON'T HARM HUMANS OR EAT SHIPS (UNLESS PROVOKED)

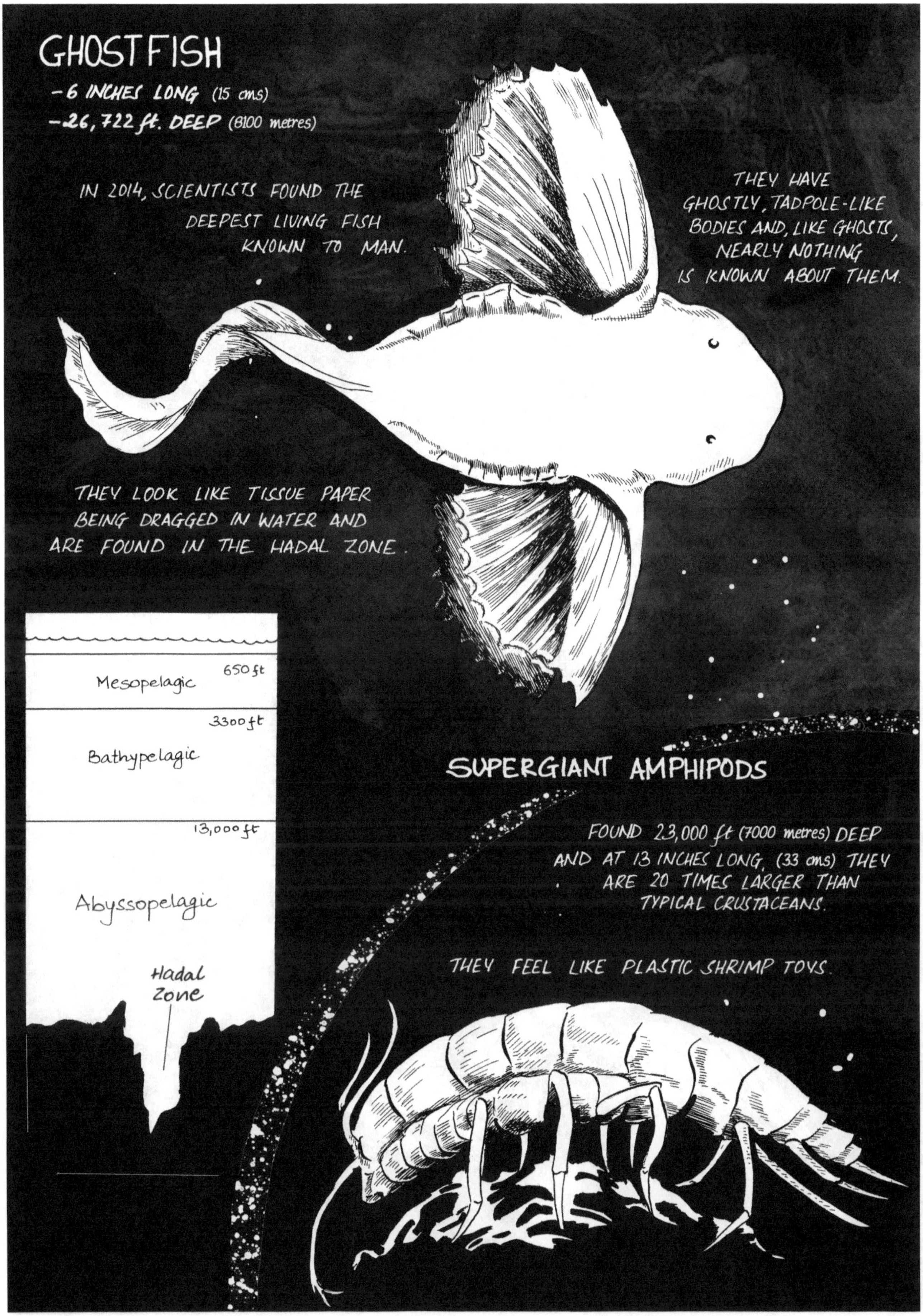
GHOSTFISH
-6 INCHES LONG (15 cms)
-26,722 ft. DEEP (8100 metres)
IN 2014, SCIENTISTS FOUND THE DEEPEST LIVING FISH KNOWN TO MAN.
THEY HAVE GHOSTLY, TADPOLE-LIKE BODIES AND, LIKE GHOSTS, NEARLY NOTHING IS KNOWN ABOUT THEM.
THEY LOOK LIKE TISSUE PAPER BEING DRAGGED IN WATER AND ARE FOUND IN THE HADAL ZONE.
650 ft
Mesopelagic
3300 ft
Bathypelagic
13,000 ft
Abyssopelagic
Hadal Zone
SUPERGIANT AMPHIPODS
FOUND 23,000 ft (7000 metres) DEEP AND AT 13 INCHES LONG, (33 cms) THEY ARE 20 TIMES LARGER THAN TYPICAL CRUSTACEANS.
THEY FEEL LIKE PLASTIC SHRIMP TOYS.

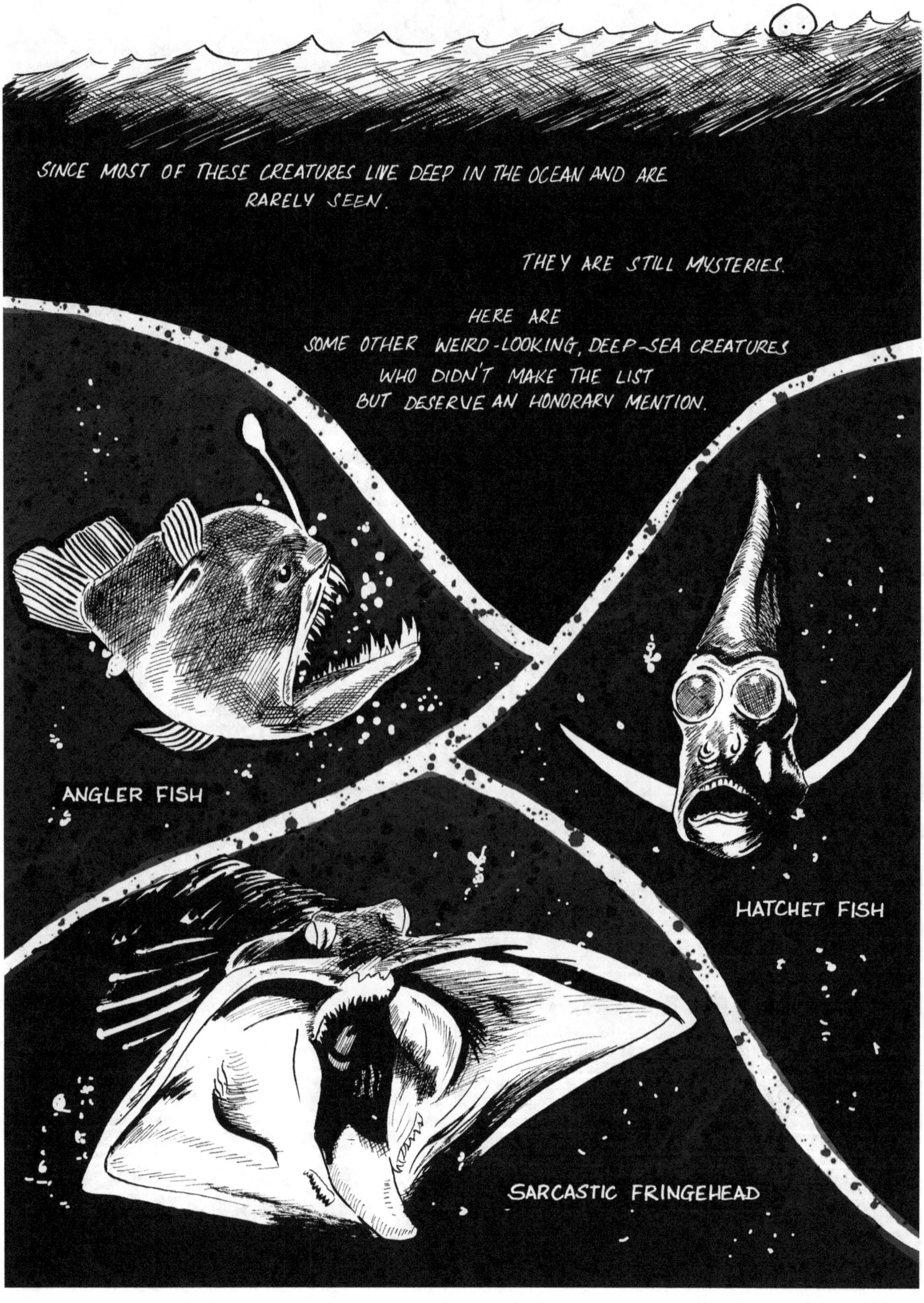
SINCE MOST OF THESE CREATURES LIVE DEEP IN THE OCEAN AND ARE RARELY SEEN.
THEY ARE STILL MYSTERIES.
HERE ARE SOME OTHER WEIRD-LOOKING, DEEP-SEA CREATURES WHO DIDN'T MAKE THE LIST BUT DESERVE AN HONORARY MENTION.
ANGLER FISH
HATCHET FISH
SARCASTIC FRINGEHEAD

Bats in Flight
By Tom O'Brien
Bats , or Chiroptera, are often considered pests, and sometimes they can be, but they're also one of the most amazing flyers in the world.
Bats are the only mammal to have true powered flight.

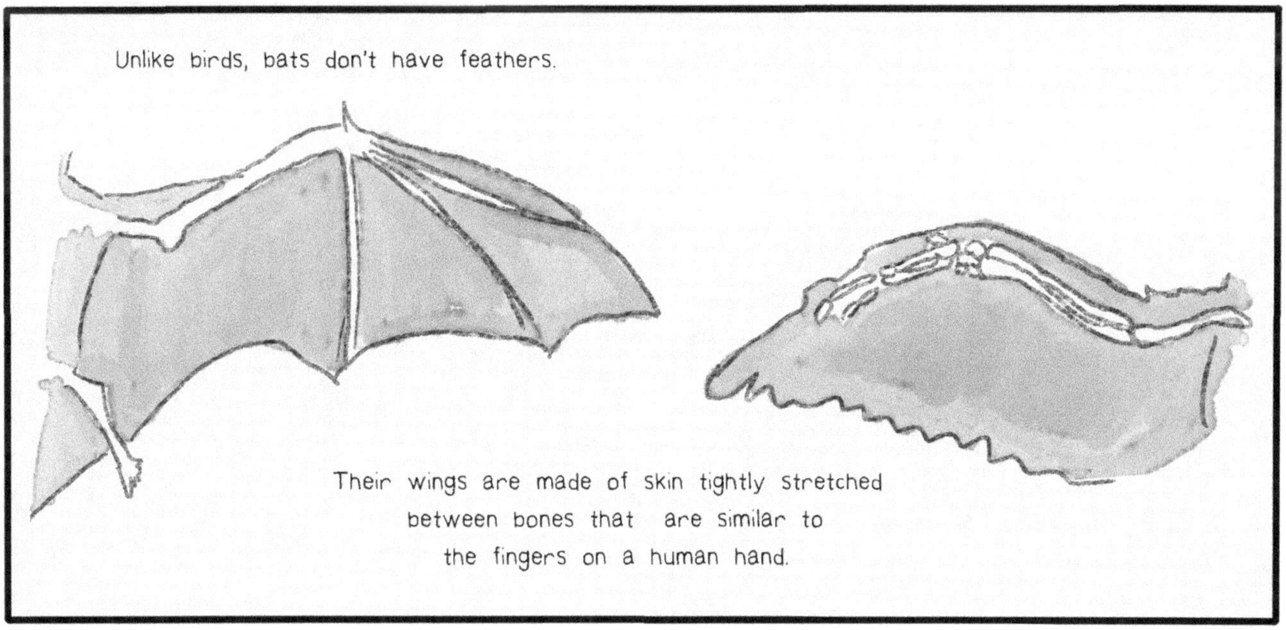
Unlike birds, bats don't have feathers.
Their wings are made of skin tightly stretched between bones that are similar to the fingers on a human hand.

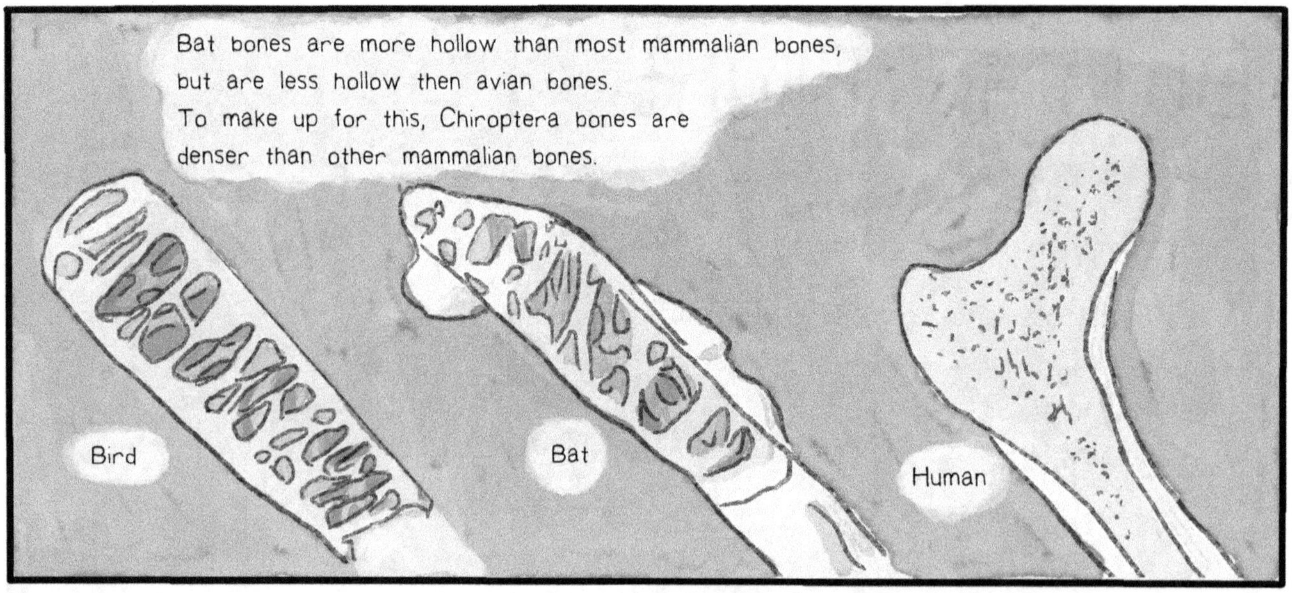
Bat bones are more hollow than most mammalian bones, but are less hollow then avian bones.
To make up for this, Chiroptera bones are denser than other mammalian bones.
Bird
Bat
Human

Despite their appearance, and their nickname "rat with wings," bats are not very closely related to mice or rats.
Some data indicates that bats' closest living relative is the flying lemur, which neither flies nor is a lemur.

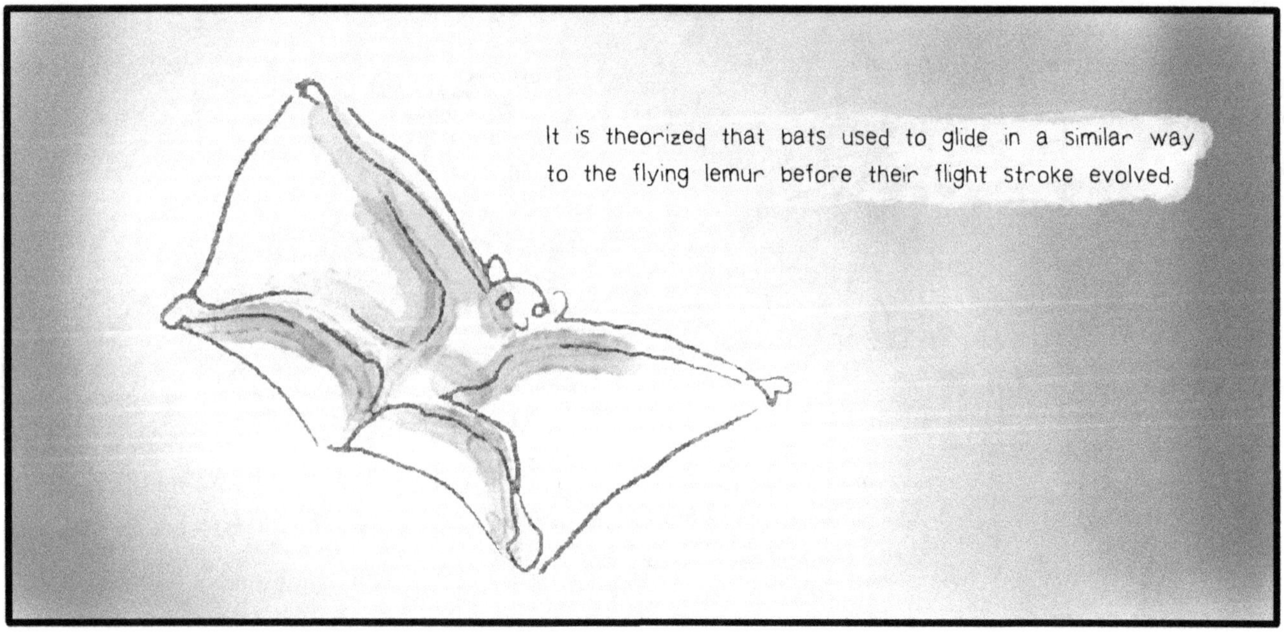
It is theorized that bats used to glide in a similar way to the flying lemur before their flight stroke evolved.

It is a widely known fact that most bats hang upside down.
This possition is necessary because most bat species, of which there are over 1,000, are incapable of taking off from the ground. This limitation is primarily due to a low flight-muscle-to-weight ratio.
The most difficult part of flying is getting up in the air.
Bats are able to skip this step by dropping into the air, rather than taking off.
This method is also of major importance to female bats, most of whom carry their young on their backs.

Bats are not solitary animals. They almost always live in groups, called colonies. Bat colonies work together in many ways.
When a large group of bats takes flight together, they fly up in a counter clockwise spiral, which creates an updraft along the column that they have formed.
This updraft makes it easier for each bat in the colony to move upward, expelling less energy to do so.

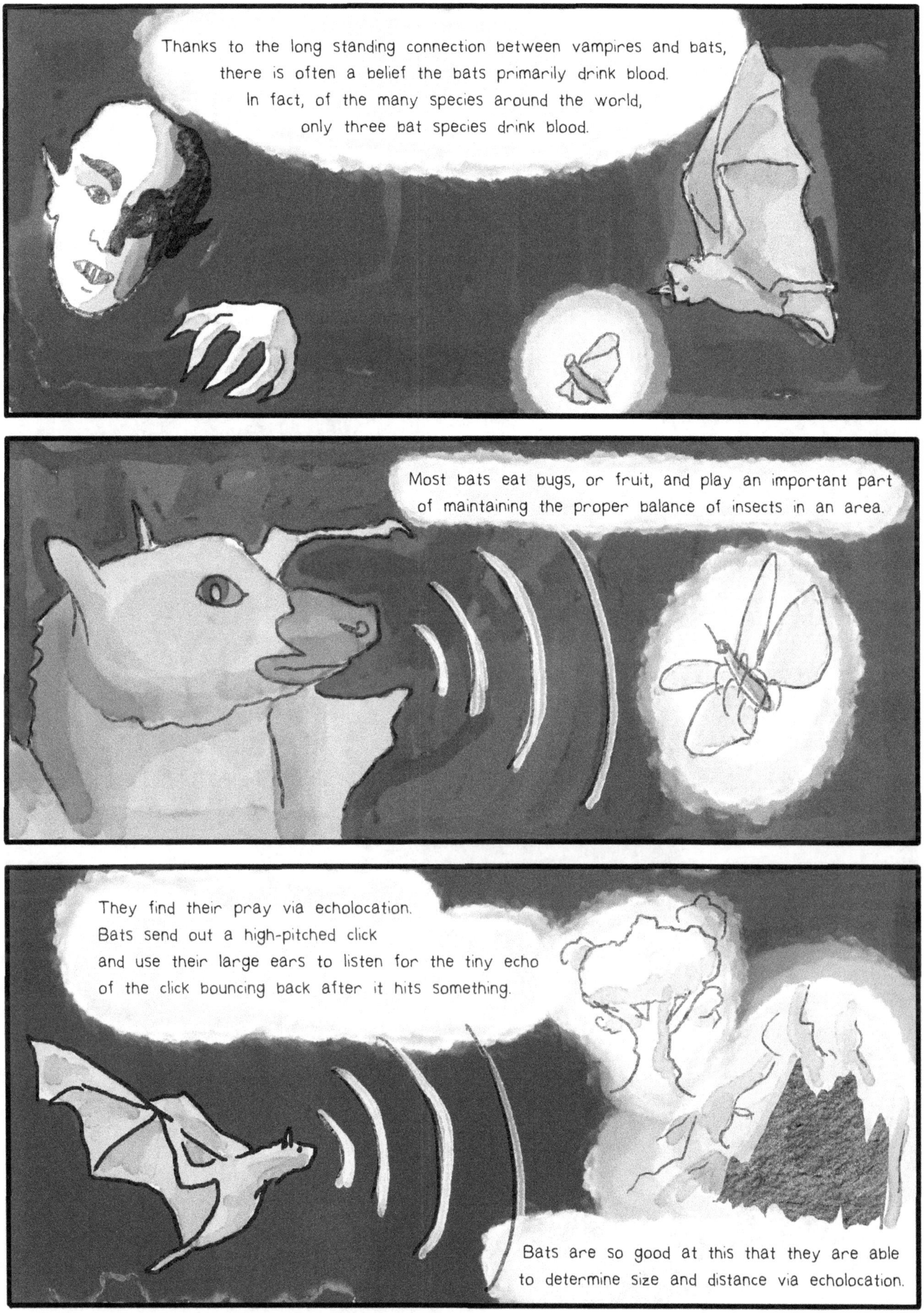
Thanks to the long standing connection between vampires and bats, there is often a belief the bats primarily drink blood. In fact, of the many species around the world, only three bat species drink blood.
Most bats eat bugs, or fruit, and play an important part of maintaining the proper balance of insects in an area.
They find their pray via echolocation. Bats send out a high-pitched click and use their large ears to listen for the tiny echo of the click bouncing back after it hits something.
Bats are so good at this that they are able to determine size and distance via echolocation.

Bats are extremely adaptable.
The survive and thrive in most environments.
Even in major cities.
Next time you're out at night, look up. Look around. There are probably bats soaring around, putting 60 million years of flight evolution to work, and they're just getting started.

RAVEN & CROW

THE DIFFERENCES BETWEEN THE COMMON RAVEN AND AMERICAN CROW

BY KELLY FERNANDEZ

UGH! YOU GUYS ALWAYS DO THIS!
IT'S JUST SO EASY TO MIX YOU TWO UP!
YEAH WELL- IT'S NOT CUTE ANYMORE!
AWW! DON'T BE LIKE THAT!
WHERE ARE YOU GOING?
I'M GONNA FIND RAVEN AND STRAIGHTEN THIS OUT!
THERE SHE IS!

WHAT'S SHE DOIN'?
SNEAK
MUNCH MUNCH
YANK
HOW MEAN!
SHE'S GOT A WEIRD SENSE OF HUMOR!
HAR HAR HAR
ZIP
YOU TWO MUST BE RELATED!
HAW HAW HAW
AND SO...
YES, IT IS TRUE!
WE BOTH BELONG TO THE CORVIDAE BIRD FAMILY, BUT IT'S UNFAIR TO MIX US UP!
I KNOW, RIGHT? WE'RE TOTALLY DIFFERENT!

FOR EXAMPLE, WE RAVENS PREFER THE WILDERNESS. YOU CAN FIND US IN TOWNS AND CITIES... BUT WE ENJOY BEING IN ISOLATED PLACES, SUCH AS THE WOODS OR MOUNTAINS.

WE ARE A LITTLE MORE SOCIAL IN OUR YOUTH. WE HELP EACH OTHER FIND FOOD AND ROOST TOGETHER AT NIGHT.

HOWEVER, WE'D RATHER BE ON OUR OWN OR IN A PAIR. AND ONCE WE SETTLE DOWN, WE DO NOT LIKE OTHER RAVENS IN OUR TERRITORY.

CROWS, ON THE OTHER HAND, AREN'T TOO CRAZY ABOUT THE FOREST. YOU CAN FIND MY BUDDIES AND ME HANGING OUT IN TOWNS, CITIES, AND OTHER URBAN SPACES.

WE'RE SUPER SOCIAL CREATURES WITH TIGHT-KNIT FAMILIES. WE STICK TOGETHER FOR YEARS AND HELP EACH OTHER RAISE CHICKS.
CAW
CAW
CAW
CAW
CAW
CAW

WE MINGLE WITH CROWS OUTSIDE OUR FAMILIES, TOO! AND WE'RE NOT AFRAID TO CHASE ANYONE OUT OF OUR TERRITORY AS AN ANGRY MOB!

OKAY, OKAY. BUT BESIDES RAVEN BEING BIGGER-
YOU TWO STILL LOOK A LOT ALIKE.
W-WELL, WE HAVE OTHER PHYSICAL DIFFERENCES!
IF YOU'D JUST LOOK A LI'L CLOSER-
YOU WILL SEE THAT RAVENS HAVE WEDGE-SHAPED TAILS AND ONLY FOUR, SLENDER FEATHERS PROTRUDING OUR WINGSPAN.
1
2
3
4
1
2
3
4
5
BUT CROWS GOT FIVE THICK FEATHERS STICKIN' OUT OUR WINGSPANS AND A SQUARE-SHAPED TAIL!

NOTICE I HAVE A LARGE, DISTINGUISHED BEAK.
WHILE I'VE GOT A NICE, MEDIUM SIZED BEAK!
ALSO, IF YOU HEAR
ROK ROK
THEN THAT WOULD BE ME!
AND IF YOU HEAR A
CAW CAW
YOU KNOW IT'S ME!
OOPS, SORRY...
NO– WE'RE SORRY!
WE SHOULD HAVE BEEN MORE CONSIDERATE TO YOU, CROW!

THANKS, YOU GUYS!
AND YOU, TOO, RAVEN. WE'RE SORRY.
O-OH! IT'S OKAY.
YOU SHOULD COME HANG OUT WITH US!
...
YEAH!
THANK YOU, BUT I'M A LONER.
FORGIVE ME. FAREWELL!
WOW
SHE'S SO COOL!
I GUESS...
AHHH
WHOOSH
WHOOSH
WOW! SOMERSAULTS! CAN YOU DO THAT, CROW?
NO! I CAN'T! WHAT A SHOW-OFF!
BEAUTIFUL!

-THE PURIFIERS-

by Kim Sucher

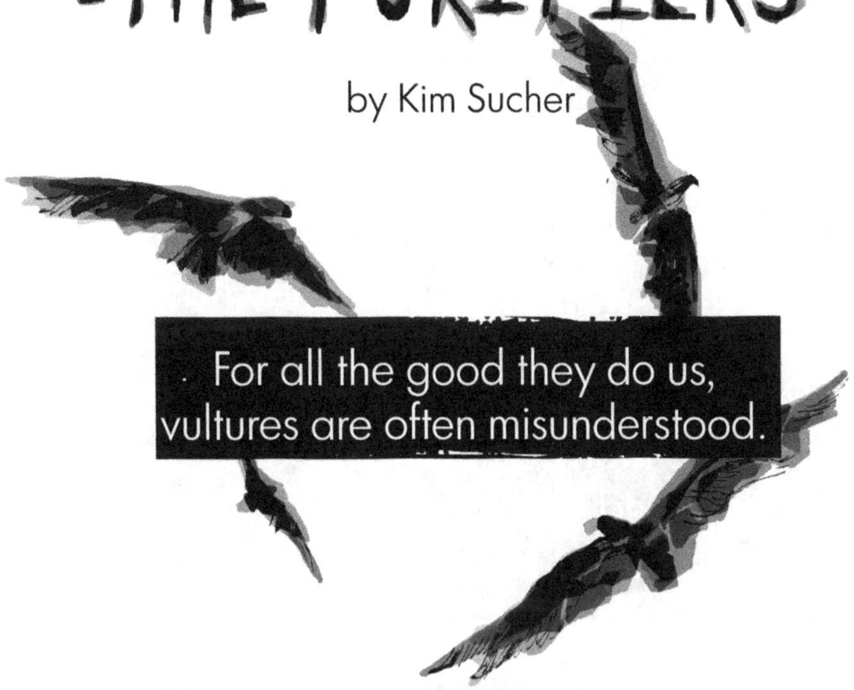

For all the good they do us, vultures are often misunderstood.

Their eating habits make people think they are dirty or disgusting.
But they can be seen as purifiers, keeping our environments clean and free of disease.

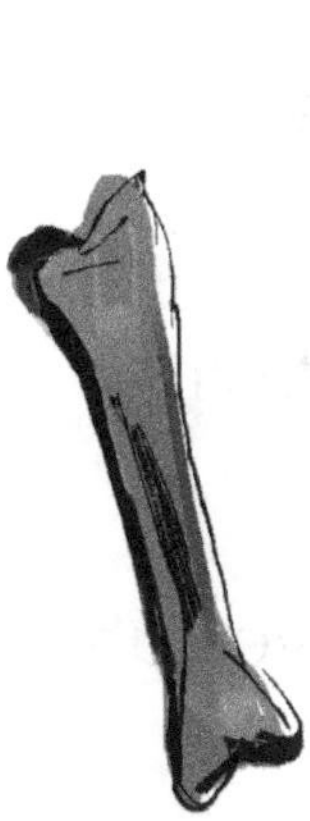

Vultures are diverse, with 23 species across our planet. They are found on every continent, except for Antarctica and Australia, and can be divided into two types: Old World and New World.

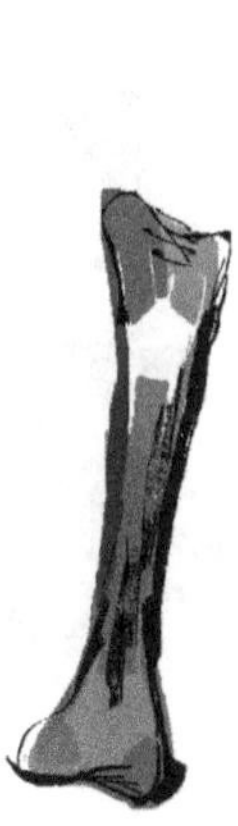

New World vs Old World

The two main categories of vultures are not closely related, but they have many similar traits because of the niche they fill in their environment. There are many differences between New and Old World vultures, though. For example:

New World

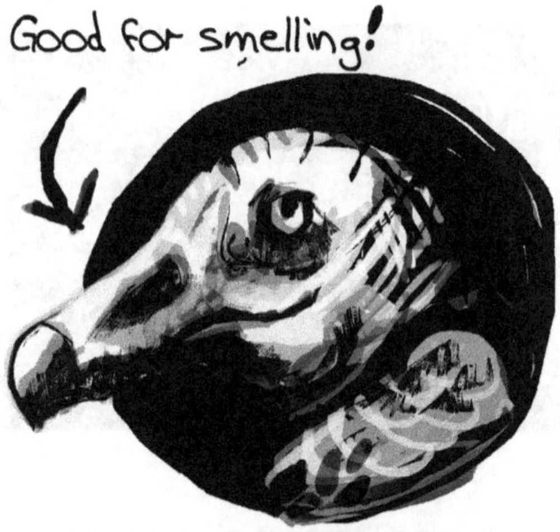

-Live in North, South, and Central America

-Have weak feet

-Locate food with a strong sense of smell

Old World

-Live in Africa, Asia, and Europe

-Have strong feet with large talons

-Mostly rely on eyesight to find food

Turkey Vulture
Cathartes aura - New World
Because of their chicken-like feet, they are usually unable to kill prey.
Their latin name, *Cathartes aura*,
means "cleansing breeze".
King Vulture
Sarcoraphus papas - New World
Ranging from Mexico to Argentina,
they can be found in many examples of ancient Mayan art.

Egyptian Vulture
Neophron percnopterus - Old World
The smallest Old World vultures!
They are one of the few animals known to use tools!
They use stones to break into large eggs.
Some have even been known to use sticks to gather wool for their nests.

Cinereous Vulture
Aegypius monachus - Old World
Also called "monk vultures" with their tuft of neck feathers resembling a monk's hood.
One of the largest birds of prey, with a wingspan that can reach up to 10 feet!
Bearded Vulture
Gypaetus barbatus- Old World
Their diet consists mostly of bones!
They drop the bones from incredible heights to get at the tasty marrow inside.
They're also known as "lammergeiers".

Griffon Vulture -
Gyps fulvus - Old World
aka the Eurasian Griffon
They are often involved in the Tibetan practice of sky-burials in which the dead are cleaned up by carrion-eating animals and the elements.
Typically, the body is first offered to the vultures, who can pick it clean quickly and efficiently.

Andean Condor
Vultur gryphus-
New World
California Condor
Gymnogyps californianus-
New World
We almost went extinct, but reintroduction has brought us from 2 or 3 dozen in the 1970s to 127 today!
We're the only New World vulture with a visible difference between males and females. (Males like me have a comb on their head!)
Both of these Condors are two of the biggest birds of prey on our planet, with wingspans up to around 10 feet from tip to tip!

Threats

Currently, 14 out of the 23 vulture species are considered threatened or endangered.

Where there are less vultures, diseases spread more easily.

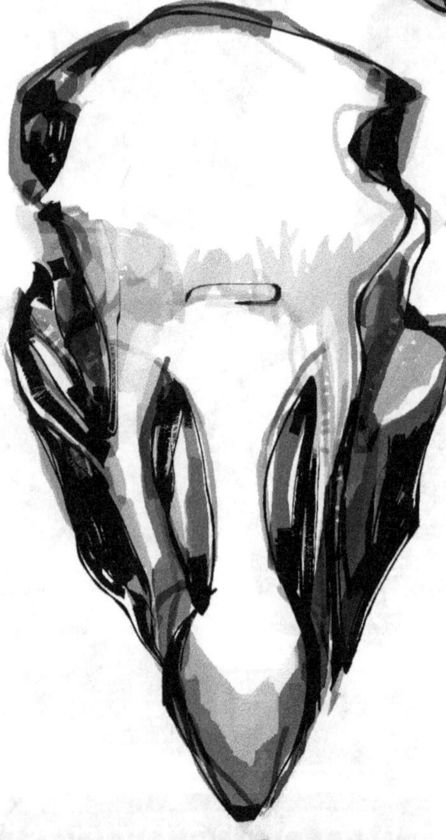

There are many factors threatening them. They can be killed by poachers or through eating poisoned livestock. In some places, they're killed for food, and in many areas, they're killed by flying into things like power lines and wind turbines.

Conservation efforts have helped save some species from the brink of extinction, but many species are still struggling!

Vultures come in many shapes and sizes, but all play the same role.
Please be kind to vultures. They help us clean our world!

Kriota Willberg

Diving in the Deep

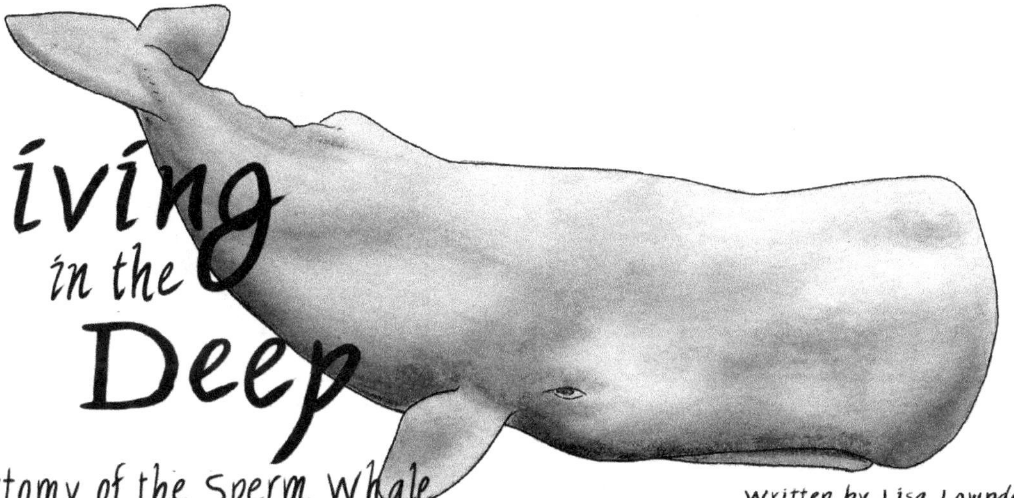

The Anatomy of the Sperm Whale

Written by Lisa Lowndes
Art by Melissa McGee

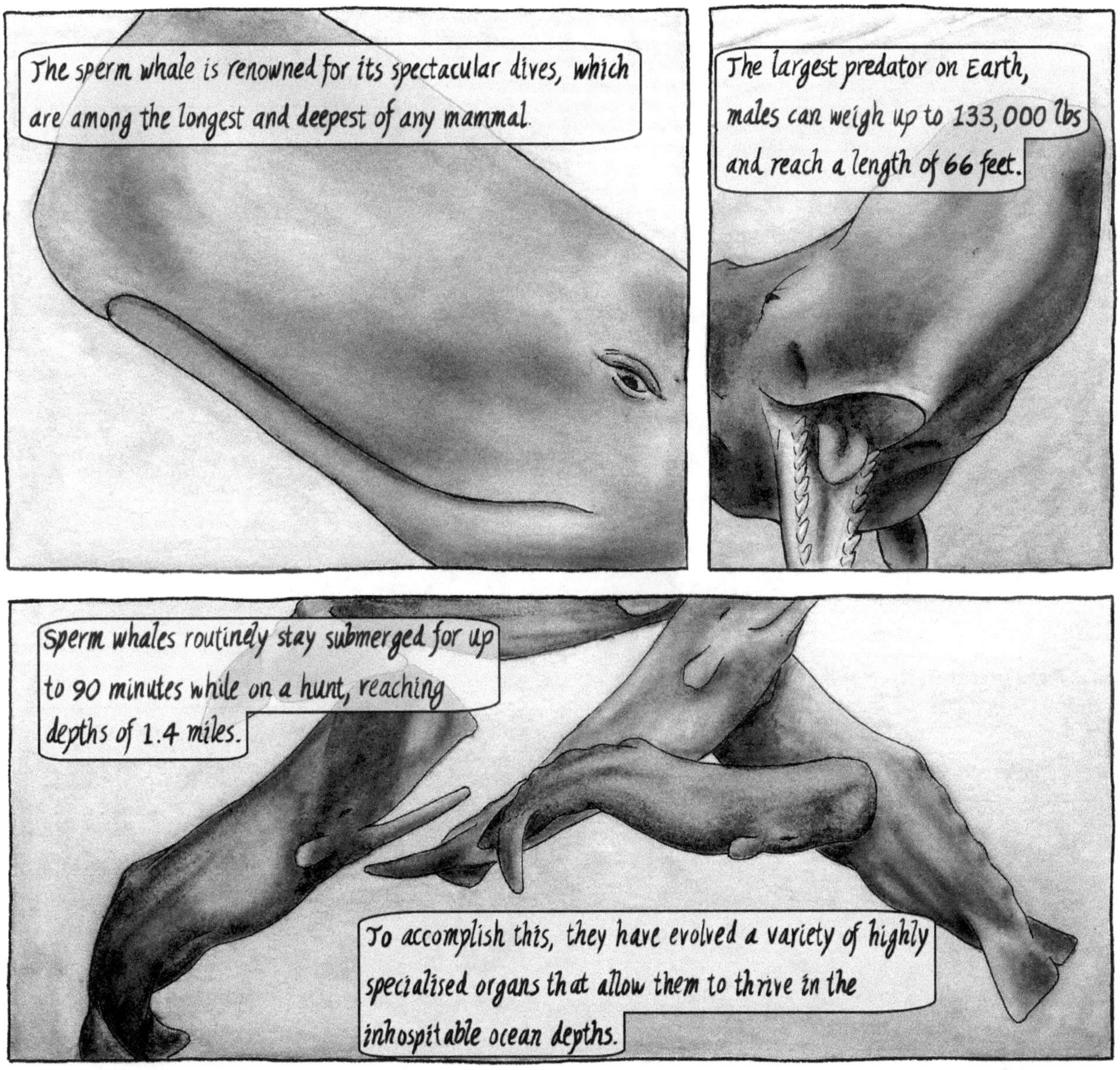

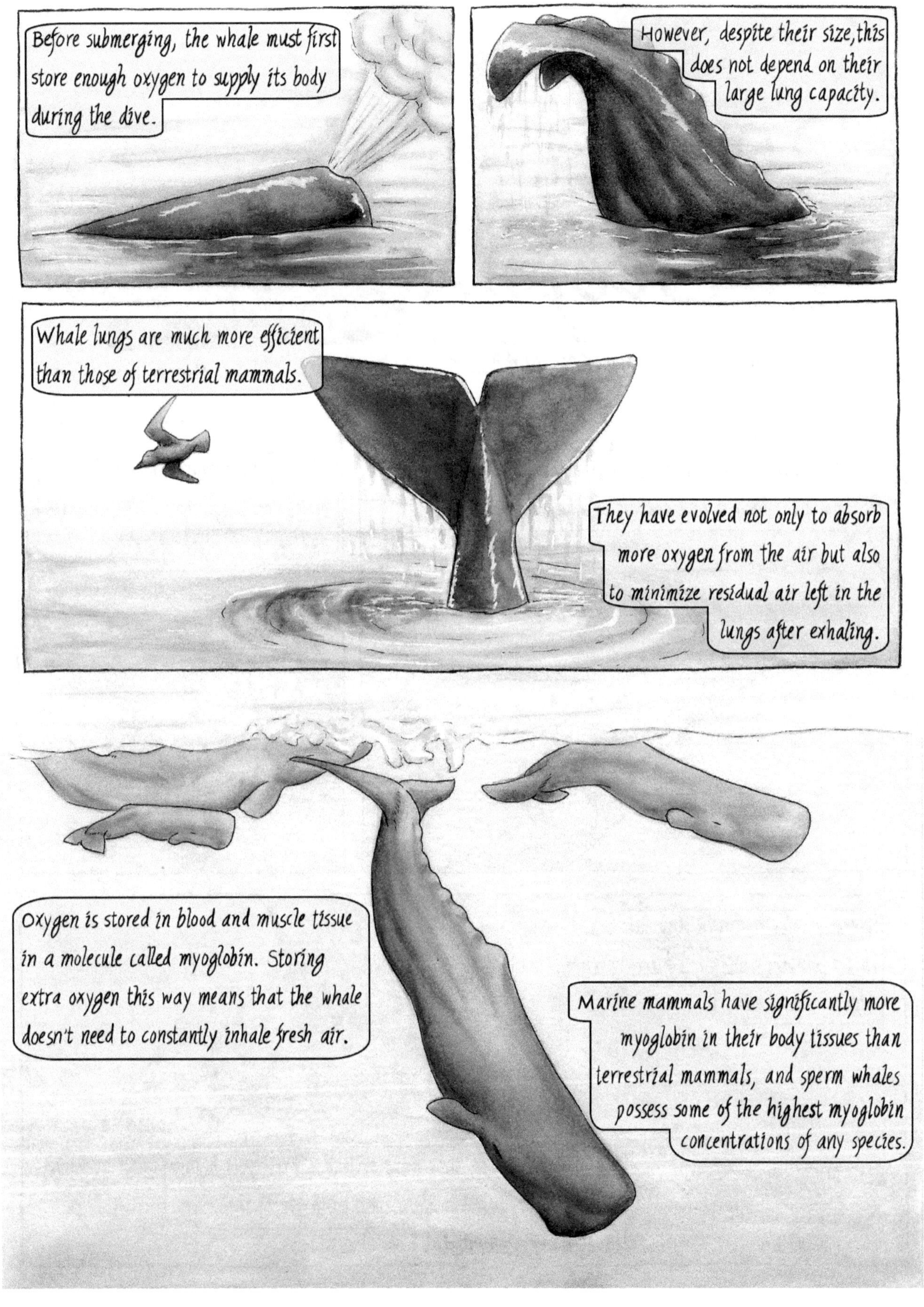
Before submerging, the whale must first store enough oxygen to supply its body during the dive.
However, despite their size, this does not depend on their large lung capacity.
Whale lungs are much more efficient than those of terrestrial mammals.
They have evolved not only to absorb more oxygen from the air but also to minimize residual air left in the lungs after exhaling.
Oxygen is stored in blood and muscle tissue in a molecule called myoglobin. Storing extra oxygen this way means that the whale doesn't need to constantly inhale fresh air.
Marine mammals have significantly more myoglobin in their body tissues than terrestrial mammals, and sperm whales possess some of the highest myoglobin concentrations of any species.

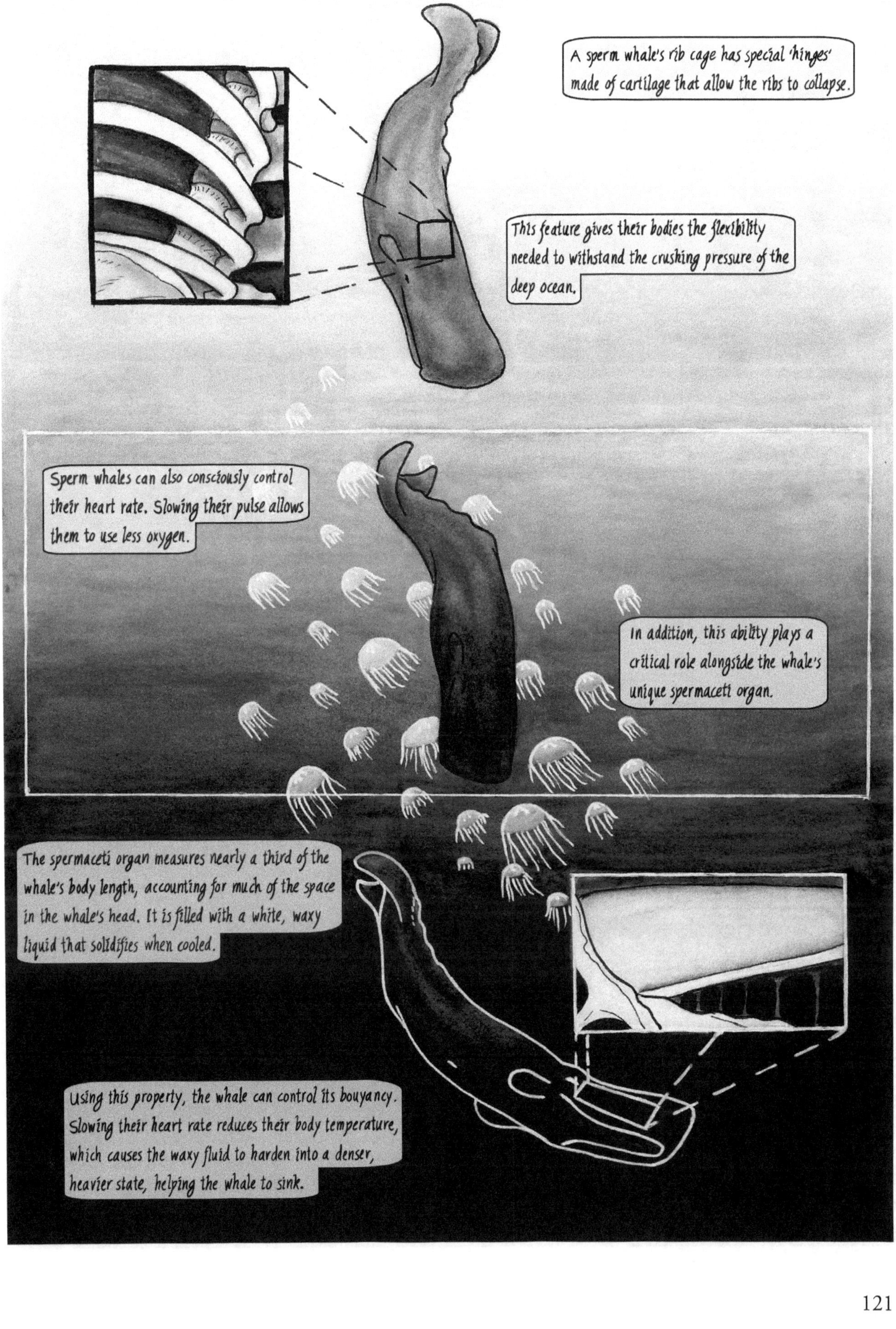
A sperm whale's rib cage has special 'hinges' made of cartilage that allow the ribs to collapse.
This feature gives their bodies the flexibility needed to withstand the crushing pressure of the deep ocean.
Sperm whales can also consciously control their heart rate. Slowing their pulse allows them to use less oxygen.
In addition, this ability plays a critical role alongside the whale's unique spermaceti organ.
The spermaceti organ measures nearly a third of the whale's body length, accounting for much of the space in the whale's head. It is filled with a white, waxy liquid that solidifies when cooled.
Using this property, the whale can control its bouyancy. Slowing their heart rate reduces their body temperature, which causes the waxy fluid to harden into a denser, heavier state, helping the whale to sink.

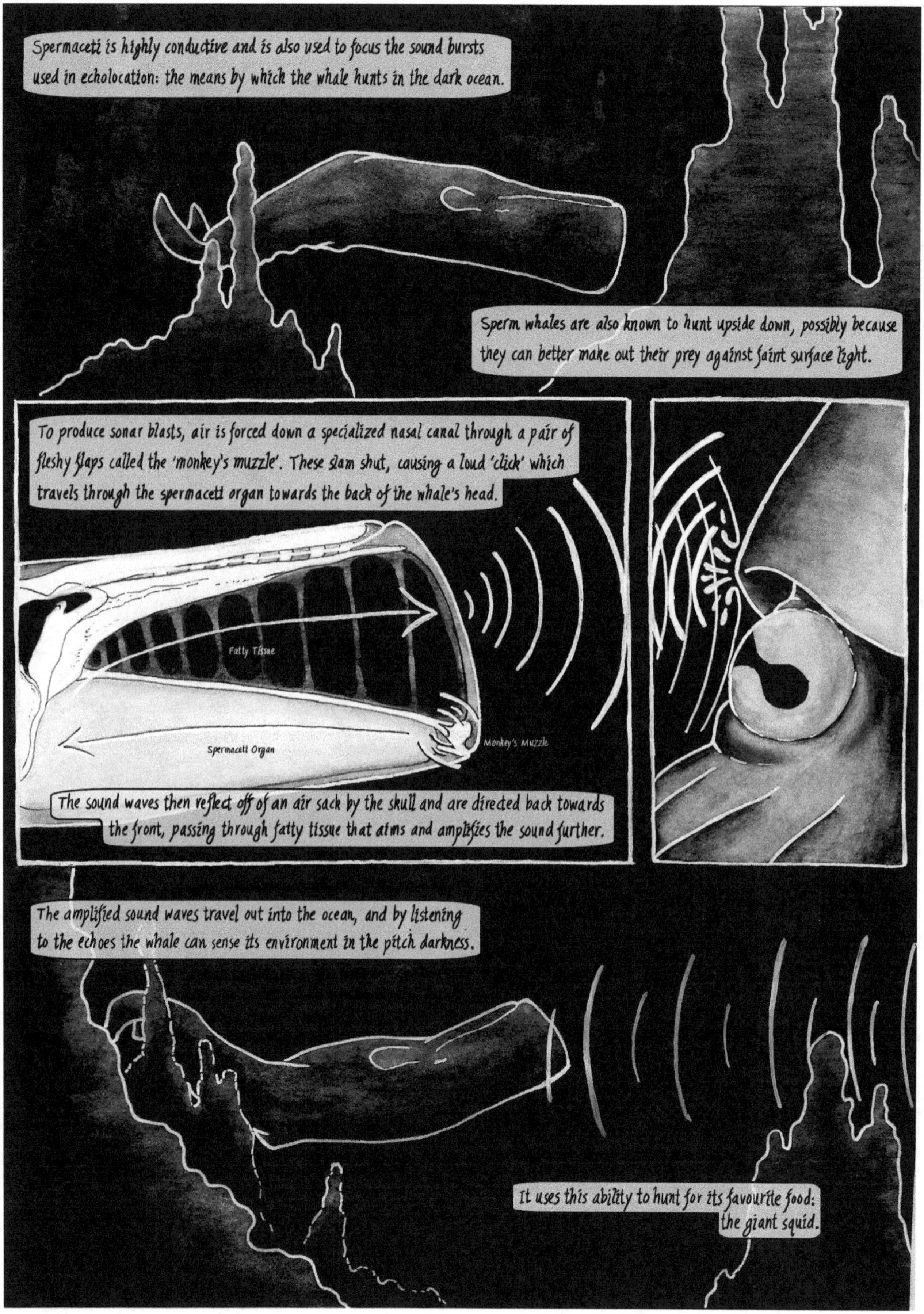
Spermaceti is highly conductive and is also used to focus the sound bursts used in echolocation: the means by which the whale hunts in the dark ocean.
Sperm whales are also known to hunt upside down, possibly because they can better make out their prey against faint surface light.
To produce sonar blasts, air is forced down a specialized nasal canal through a pair of fleshy flaps called the 'monkey's muzzle'. These slam shut, causing a loud 'click' which travels through the spermaceti organ towards the back of the whale's head.
Fatty Tissue
Spermaceti Organ
Monkey's Muzzle
The sound waves then reflect off of an air sack by the skull and are directed back towards the front, passing through fatty tissue that aims and amplifies the sound further.
The amplified sound waves travel out into the ocean, and by listening to the echoes the whale can sense its environment in the pitch darkness.
It uses this ability to hunt for its favourite food: the giant squid.

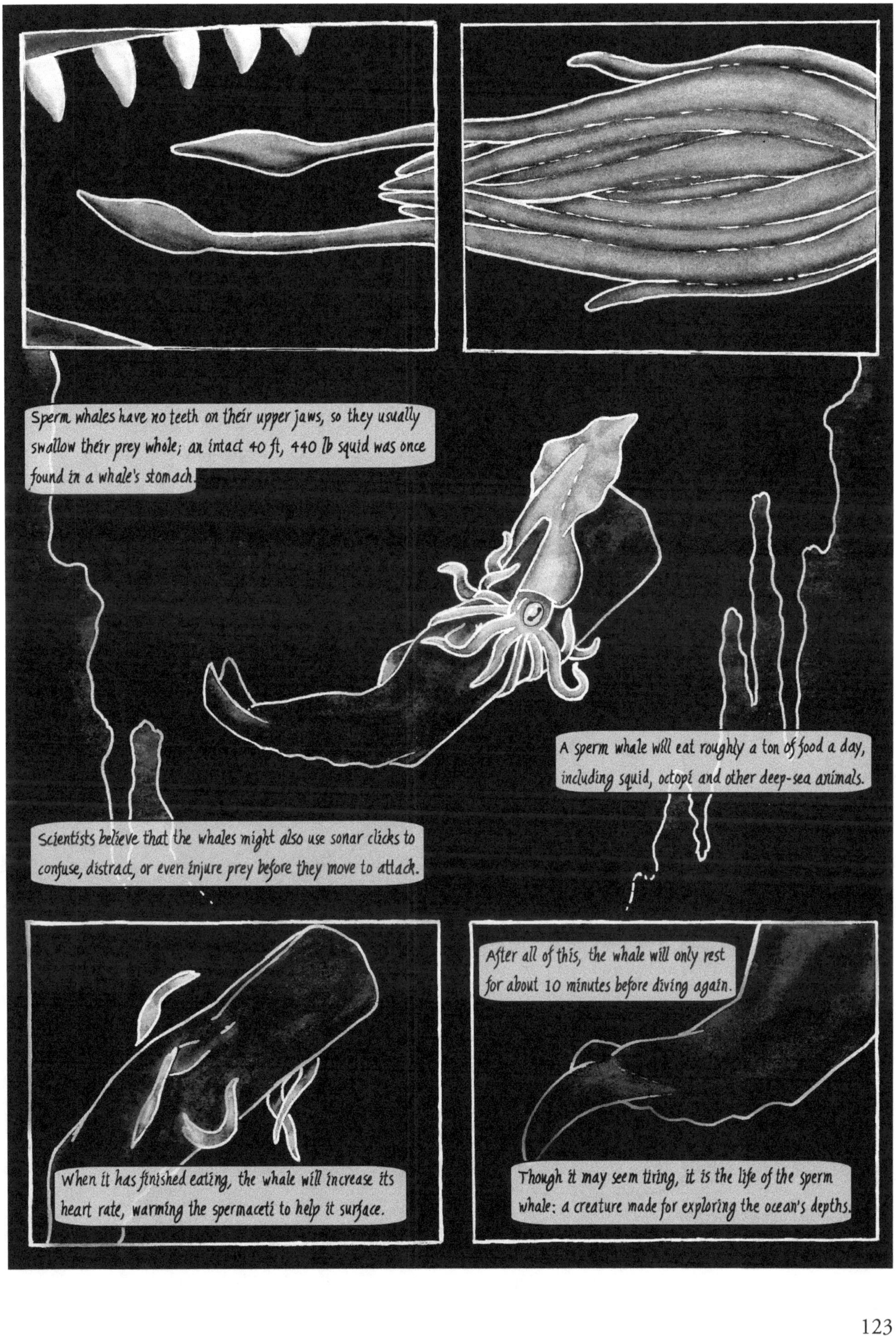
Sperm whales have no teeth on their upper jaws, so they usually swallow their prey whole; an intact 40 ft, 440 lb squid was once found in a whale's stomach.
A sperm whale will eat roughly a ton of food a day, including squid, octopi and other deep-sea animals.
Scientists believe that the whales might also use sonar clicks to confuse, distract, or even injure prey before they move to attack.
When it has finished eating, the whale will increase its heart rate, warming the spermaceti to help it surface.
After all of this, the whale will only rest for about 10 minutes before diving again.
Though it may seem tiring, it is the life of the sperm whale: a creature made for exploring the ocean's depths.

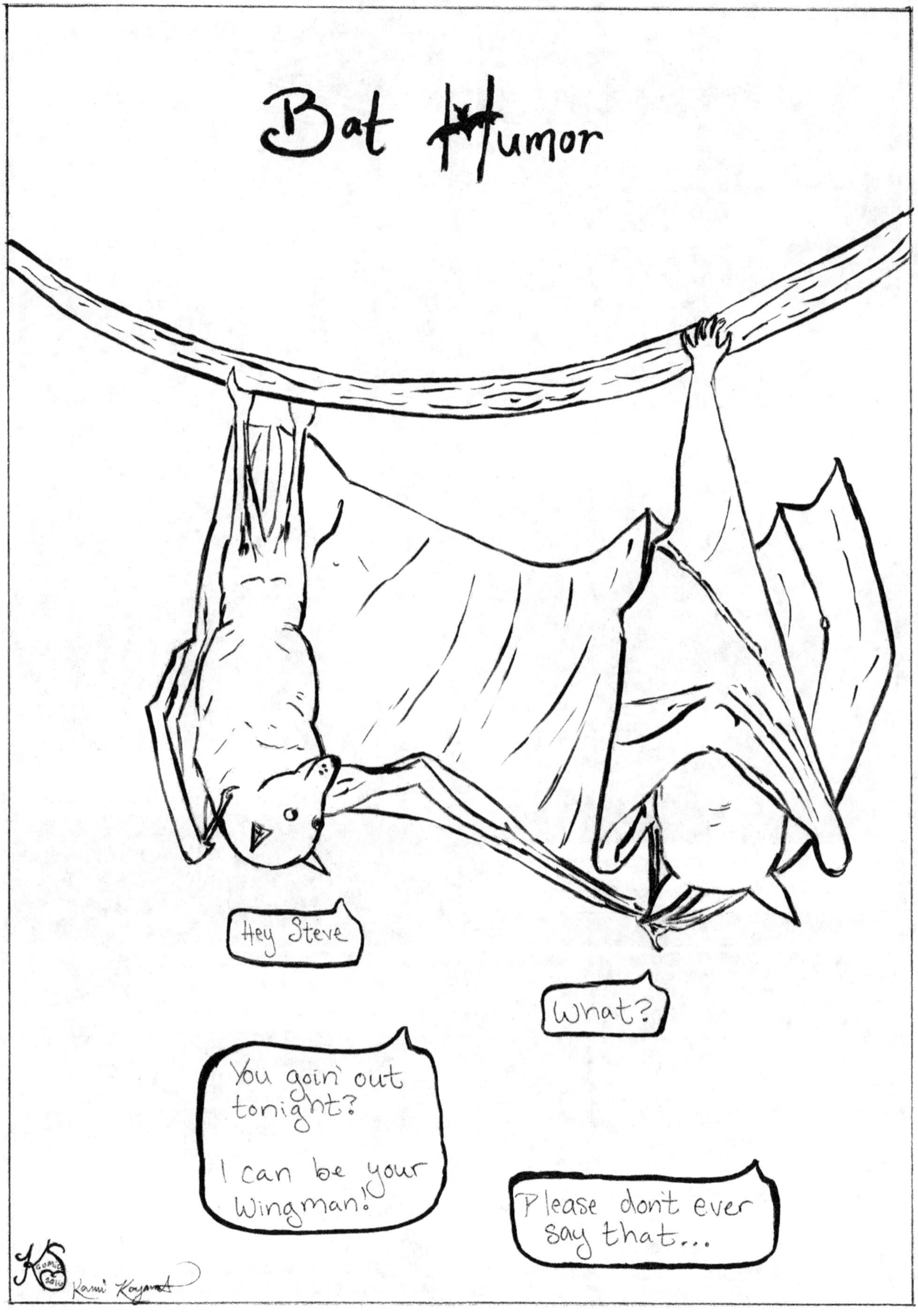
Bat Humor
Hey Steve
What?
You goin' out tonight?
I can be your Wingman!
Please don't ever say that...
Kami Koyama

There are, as no doubt you already know, many fish in the sea.
UNDERWATER TAILS
Kelly Swann
You humans estimate that there may be over 30,000 species of fish in the sea, having found just over 15,000 so far.
And you've found 47 species of those in my genus so far...

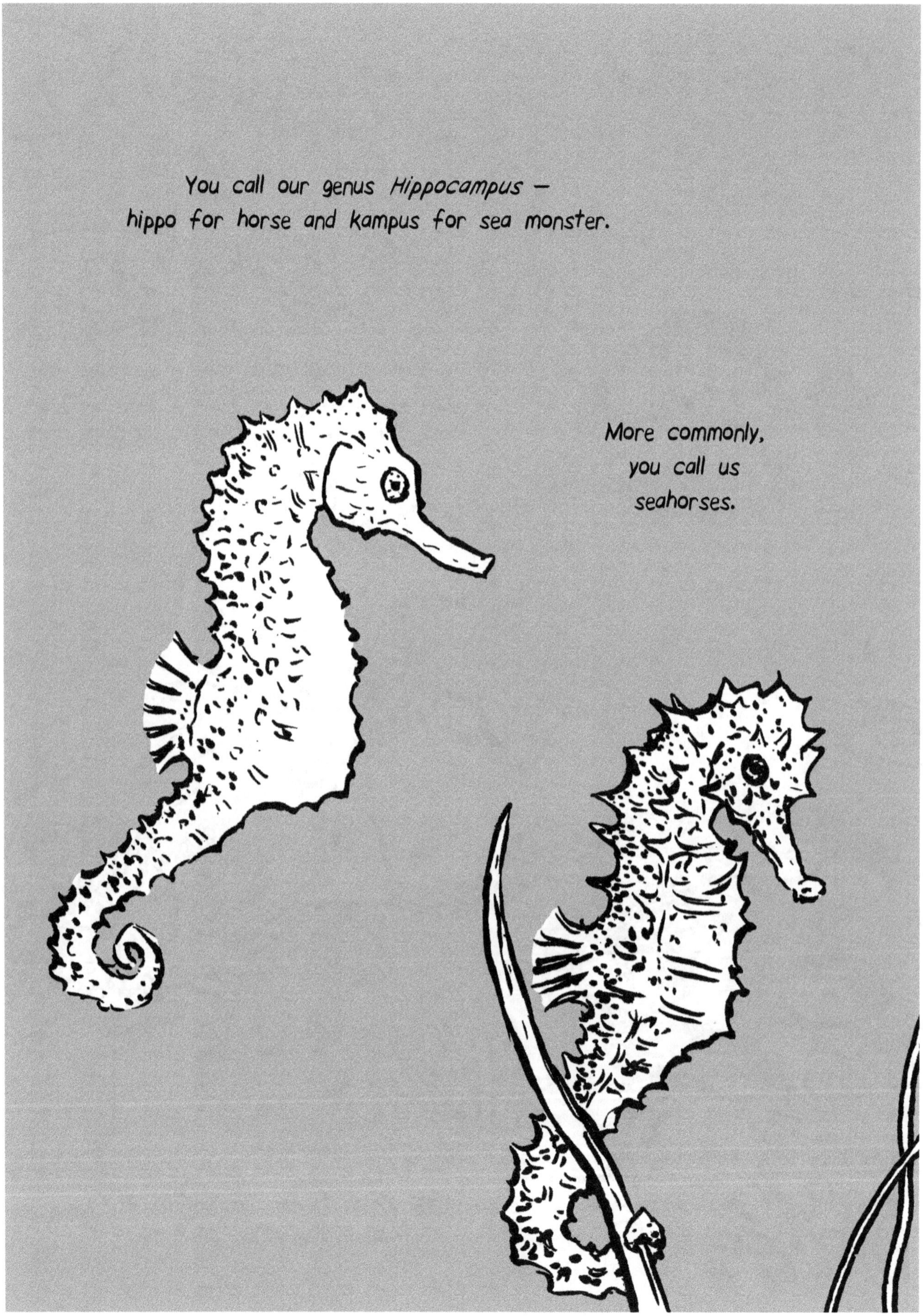
You call our genus *Hippocampus* —
hippo for horse and kampus for sea monster.
More commonly,
you call us
seahorses.

And here we are.
These fry (baby seahorses) are my brothers and sisters.
And this is me.
We were born just minutes ago, and now we're on our own, at least in a manner of speaking.

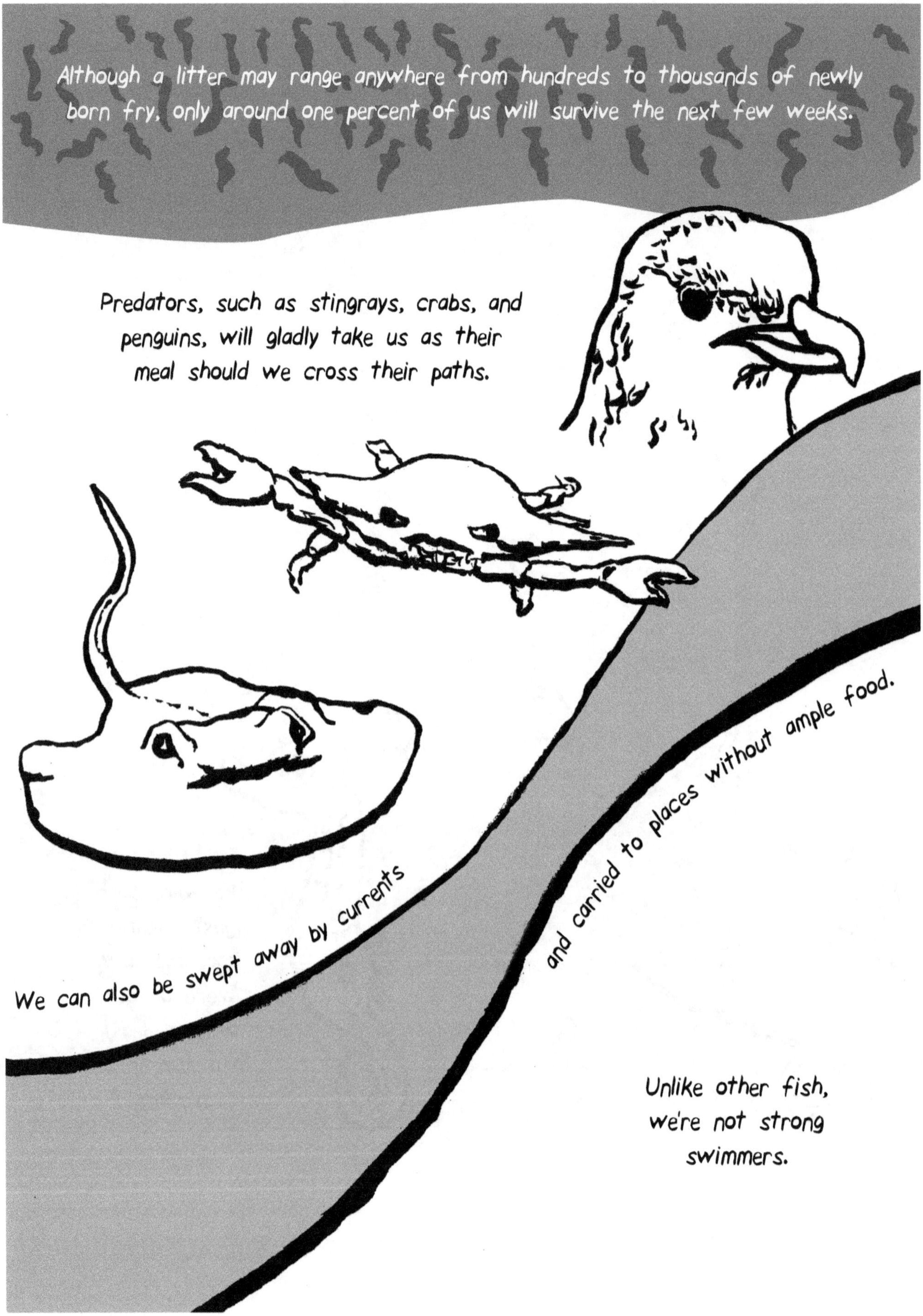
Although a litter may range anywhere from hundreds to thousands of newly born fry, only around one percent of us will survive the next few weeks.
Predators, such as stingrays, crabs, and penguins, will gladly take us as their meal should we cross their paths.
We can also be swept away by currents
and carried to places without ample food.
Unlike other fish, we're not strong swimmers.

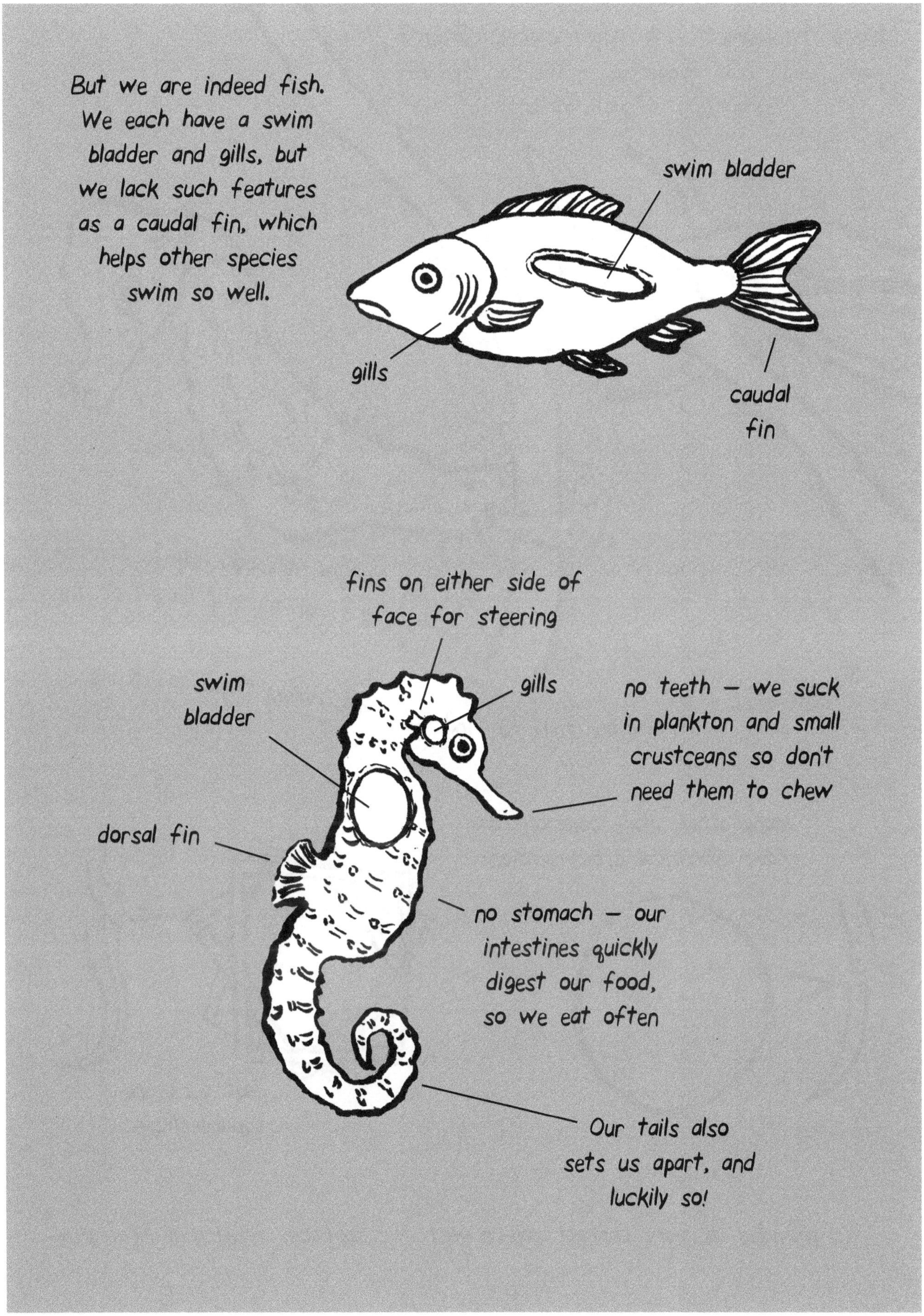
But we are indeed fish. We each have a swim bladder and gills, but we lack such features as a caudal fin, which helps other species swim so well.
swim bladder
gills
caudal fin
fins on either side of face for steering
swim bladder
gills
no teeth – we suck in plankton and small crustceans so don't need them to chew
dorsal fin
no stomach – our intestines quickly digest our food, so we eat often
Our tails also sets us apart, and luckily so!

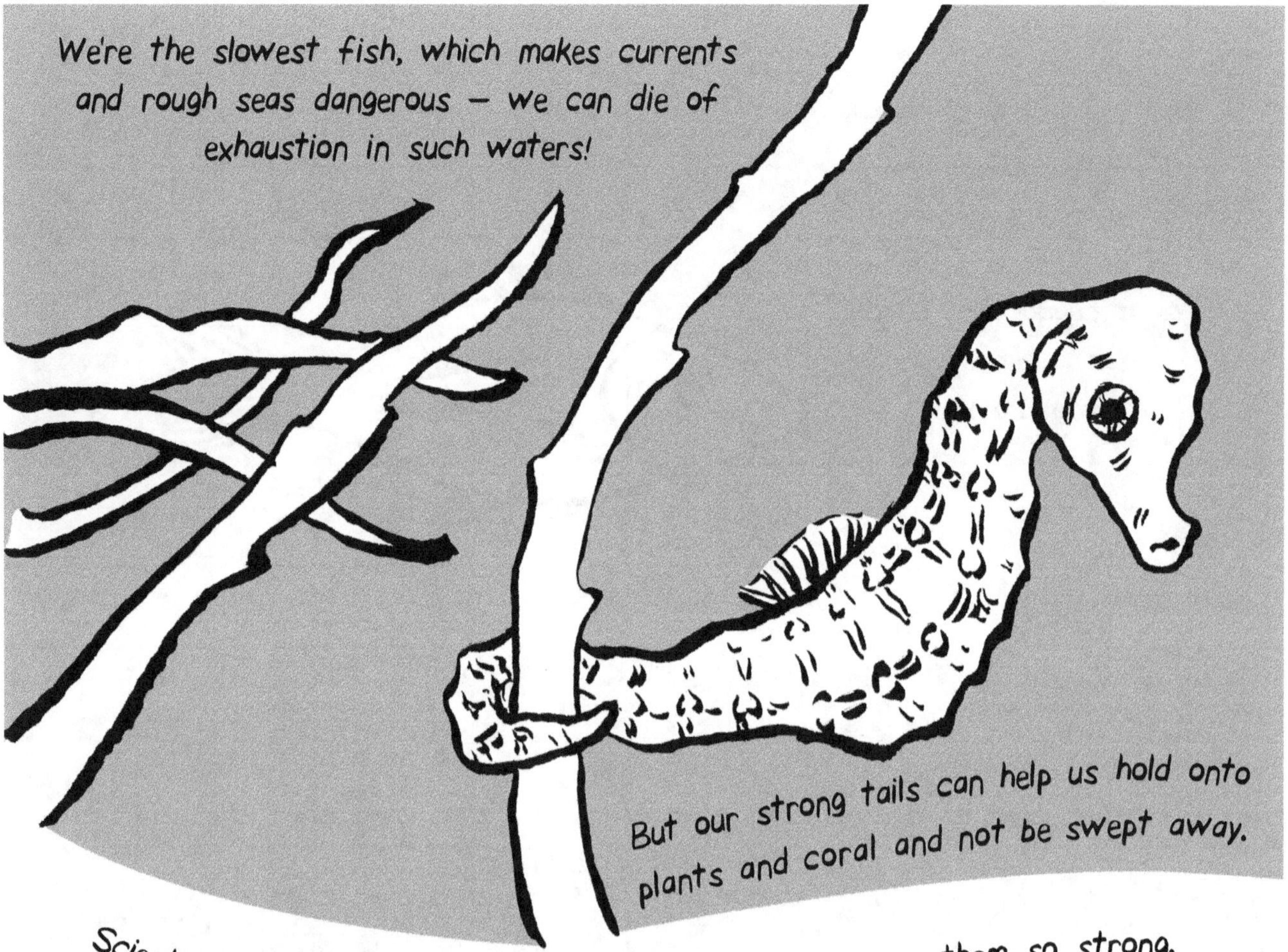

Scientists have studied our tails to find out what makes them so strong.

Many other, more common tails have cylindrical cross-sections.

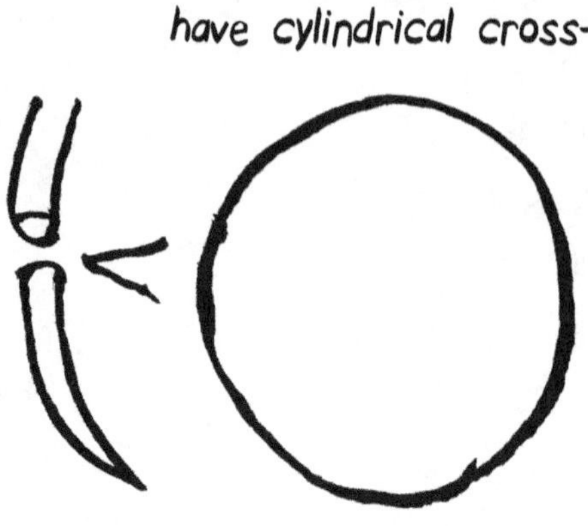

But ours are square-shaped.

This provides us more contact points with the surface, allowing a firm grip.

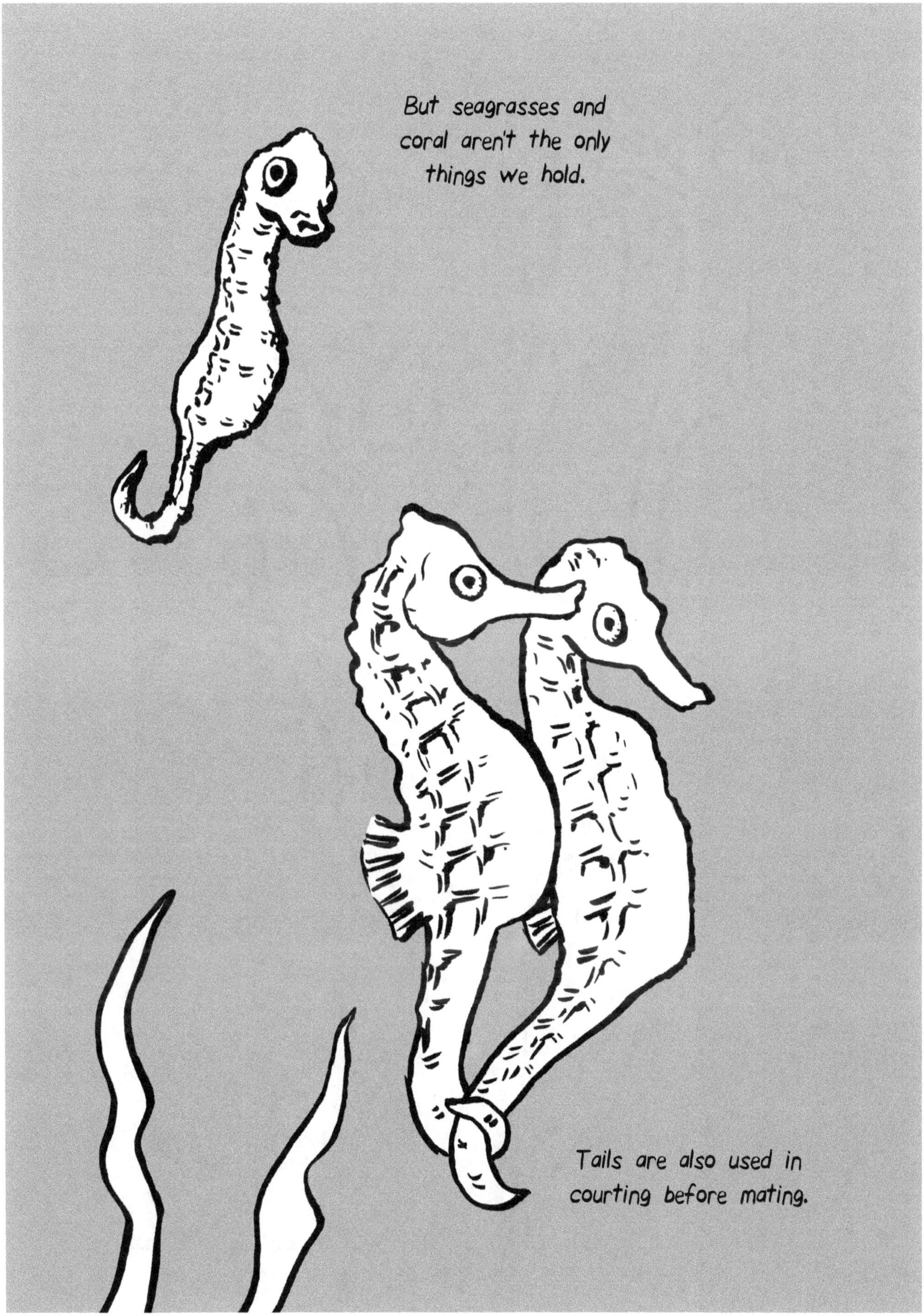
But seagrasses and
coral aren't the only
things we hold.
Tails are also used in
courting before mating.

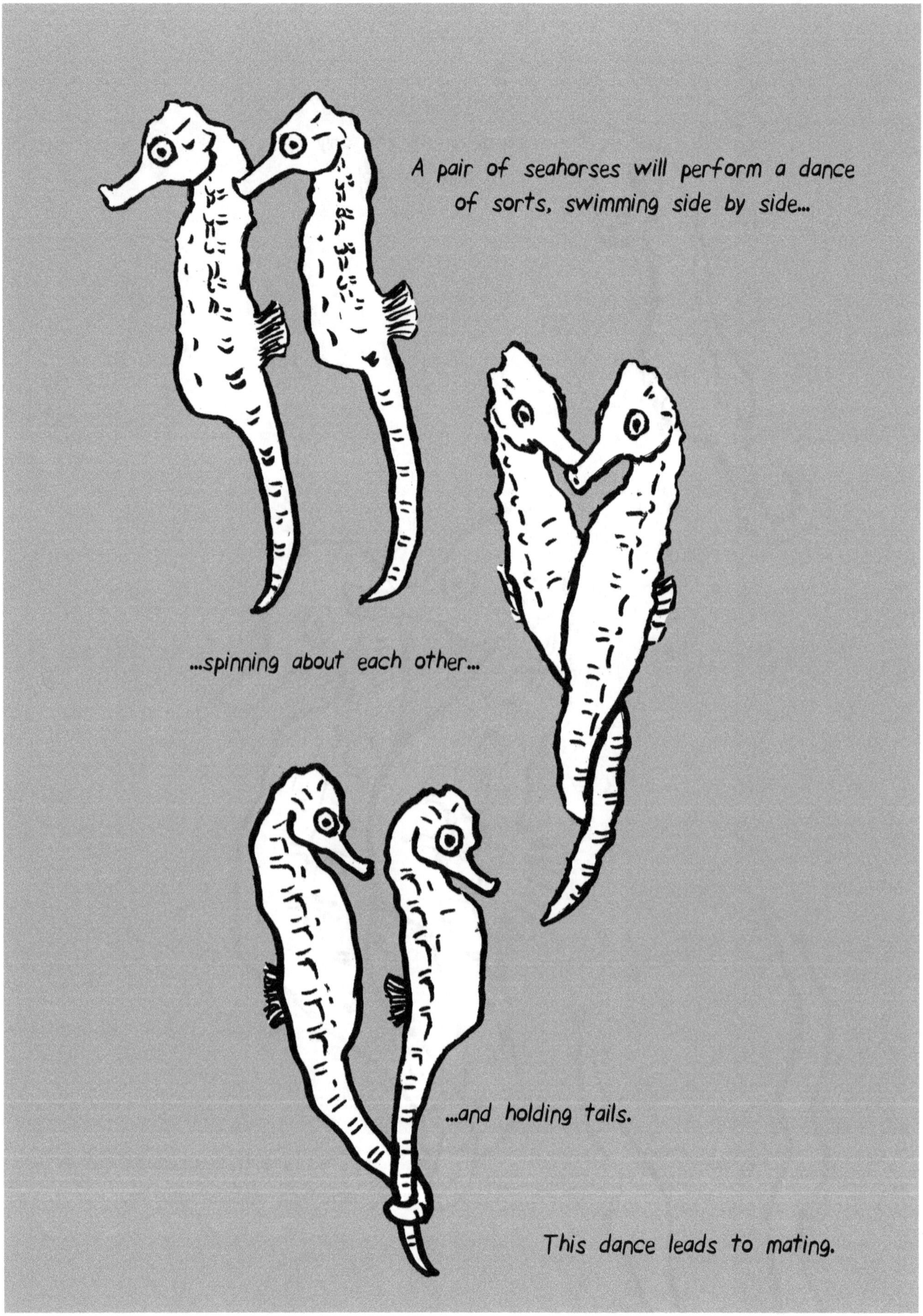
A pair of seahorses will perform a dance of sorts, swimming side by side...
...spinning about each other...
...and holding tails.
This dance leads to mating.

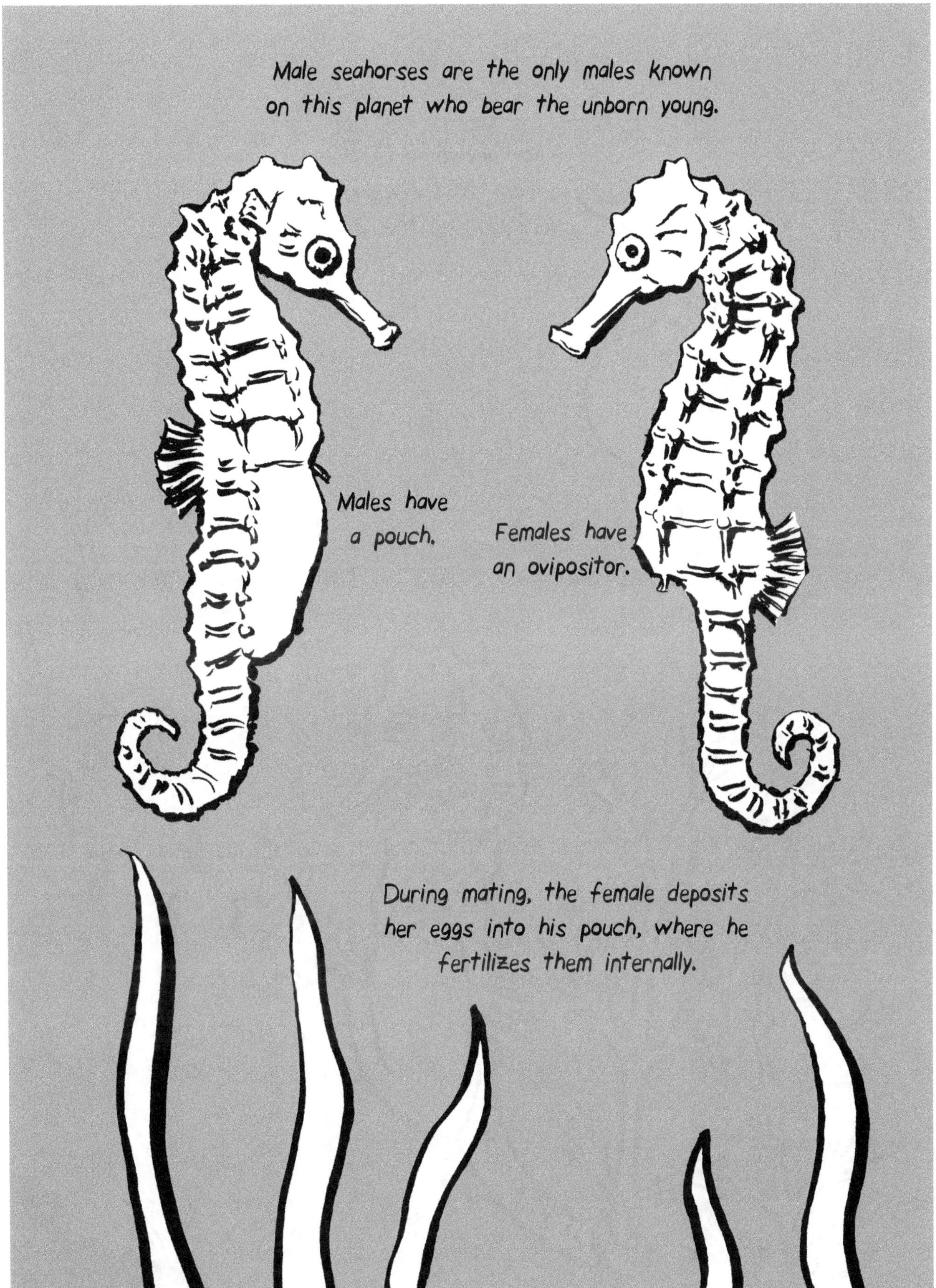
Male seahorses are the only males known on this planet who bear the unborn young.
Males have a pouch.
Females have an ovipositor.
During mating, the female deposits her eggs into his pouch, where he fertilizes them internally.

He carries the eggs
for anywhere between
one to six weeks,
depending on the species.
Then he releases the
miniature seahorses
into the ocean.

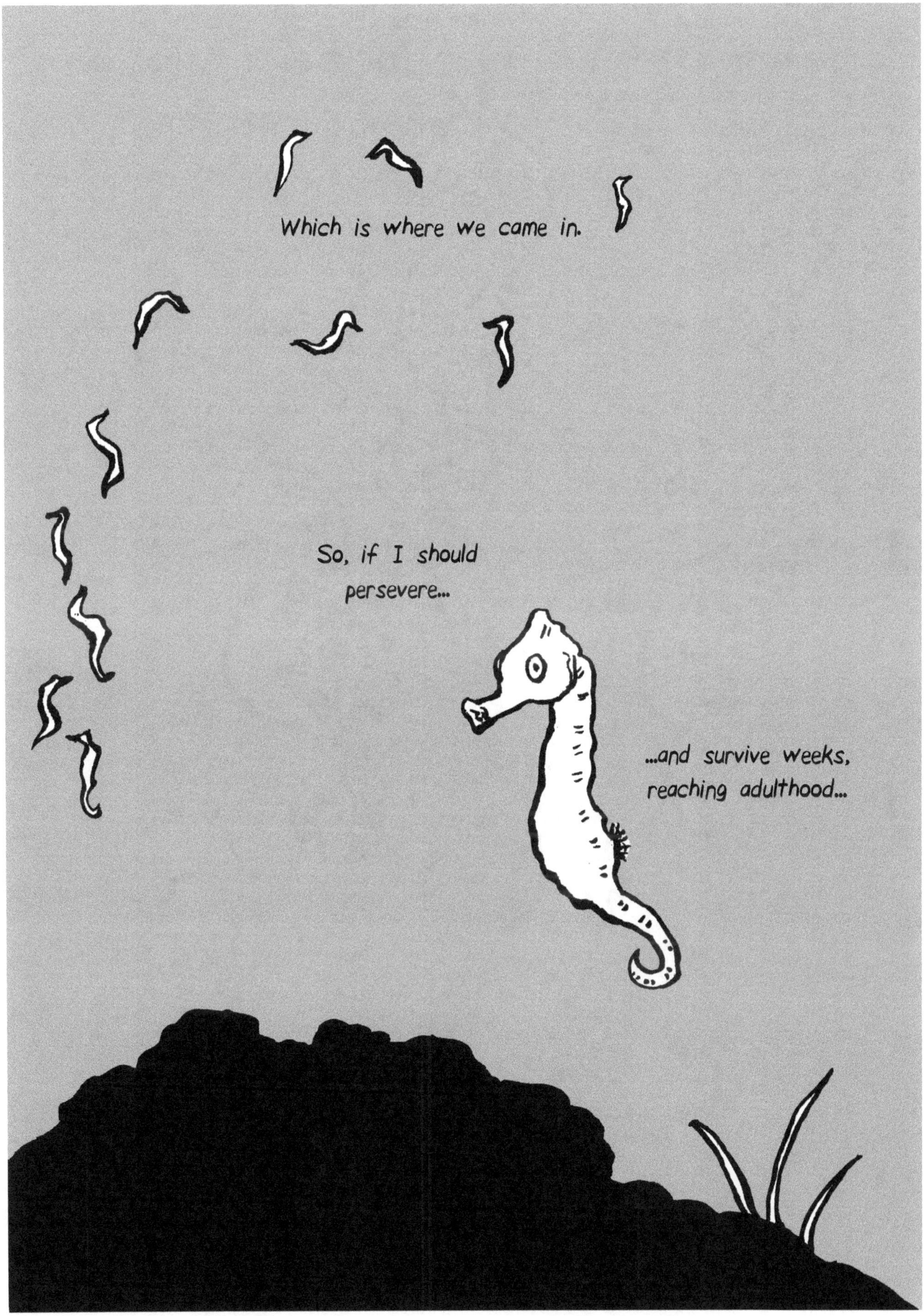
Which is where we came in.
So, if I should persevere...
...and survive weeks, reaching adulthood...

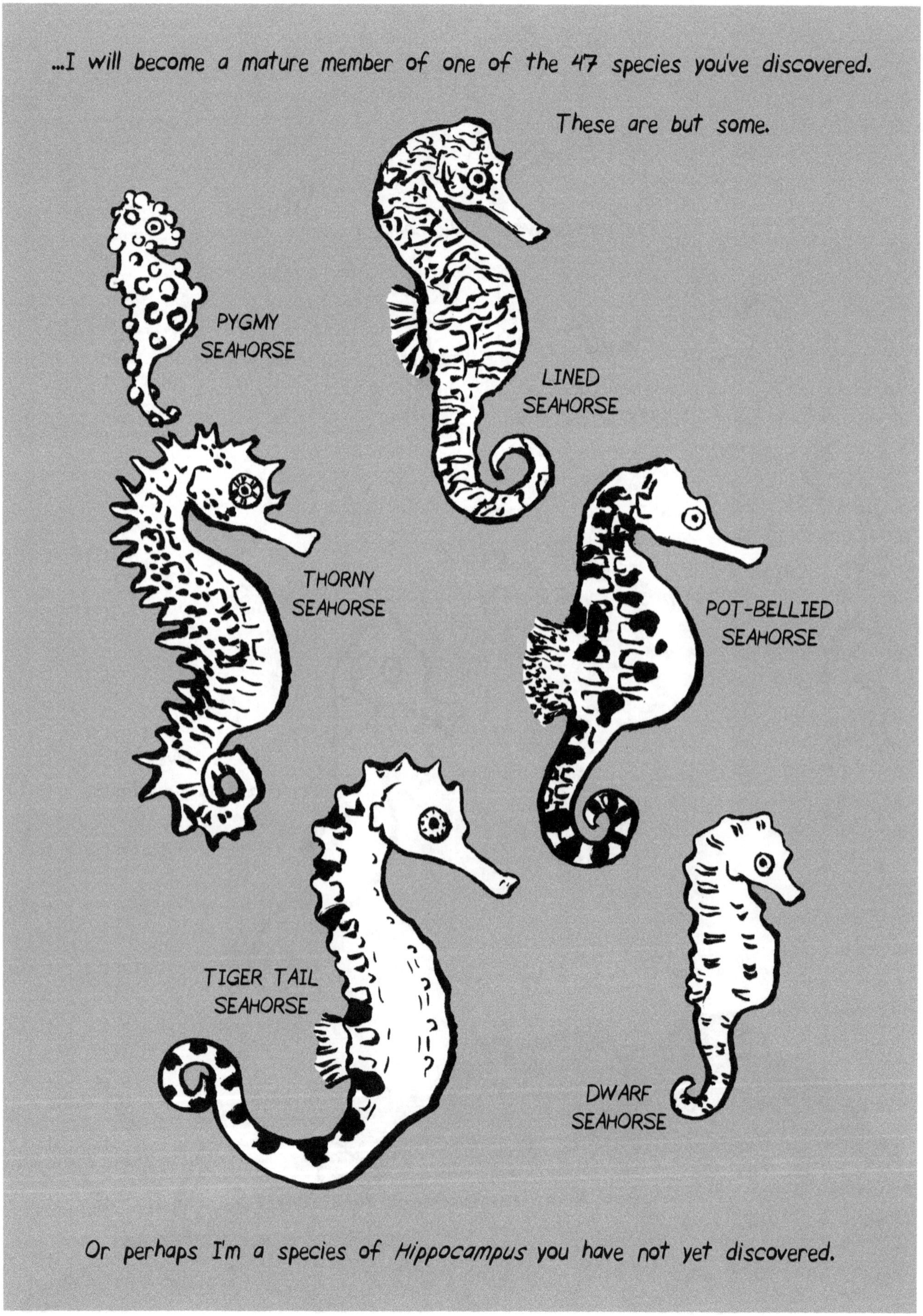
...I will become a mature member of one of the 47 species you've discovered.
These are but some.
PYGMY SEAHORSE
LINED SEAHORSE
THORNY SEAHORSE
POT-BELLIED SEAHORSE
TIGER TAIL SEAHORSE
DWARF SEAHORSE
Or perhaps I'm a species of *Hippocampus* you have not yet discovered.

This is our home, Chobe National Park, a lush and extensive marshland in the northeast of Botswana in Southern Africa.
Matriarch
by Salakjit

My name is Maya.
And this is my family.

I am their matriarch,
the leader of our herd.

We live by social rules
that are intricate,
and we have a fairly
unique family structure.
The females rule,
and it is up to the most
experienced to lead now that
our mother has passed.
Because I am the most
mature in the group,
that responsibility
falls on me.
As the backbone
of my family,
I am charged with
making decisions that
ensure my family safety,
health, and survival.

I am not just a matriarch.
I am also a mother to
my young daughter, Zena.
Her father does not live with us.
In our society, males live in
solitary or with other males
in a bachelor band.
Like other calves,
young Zena is completely
dependent on her family
to survive.
We provide her with
food, love, and discipline.

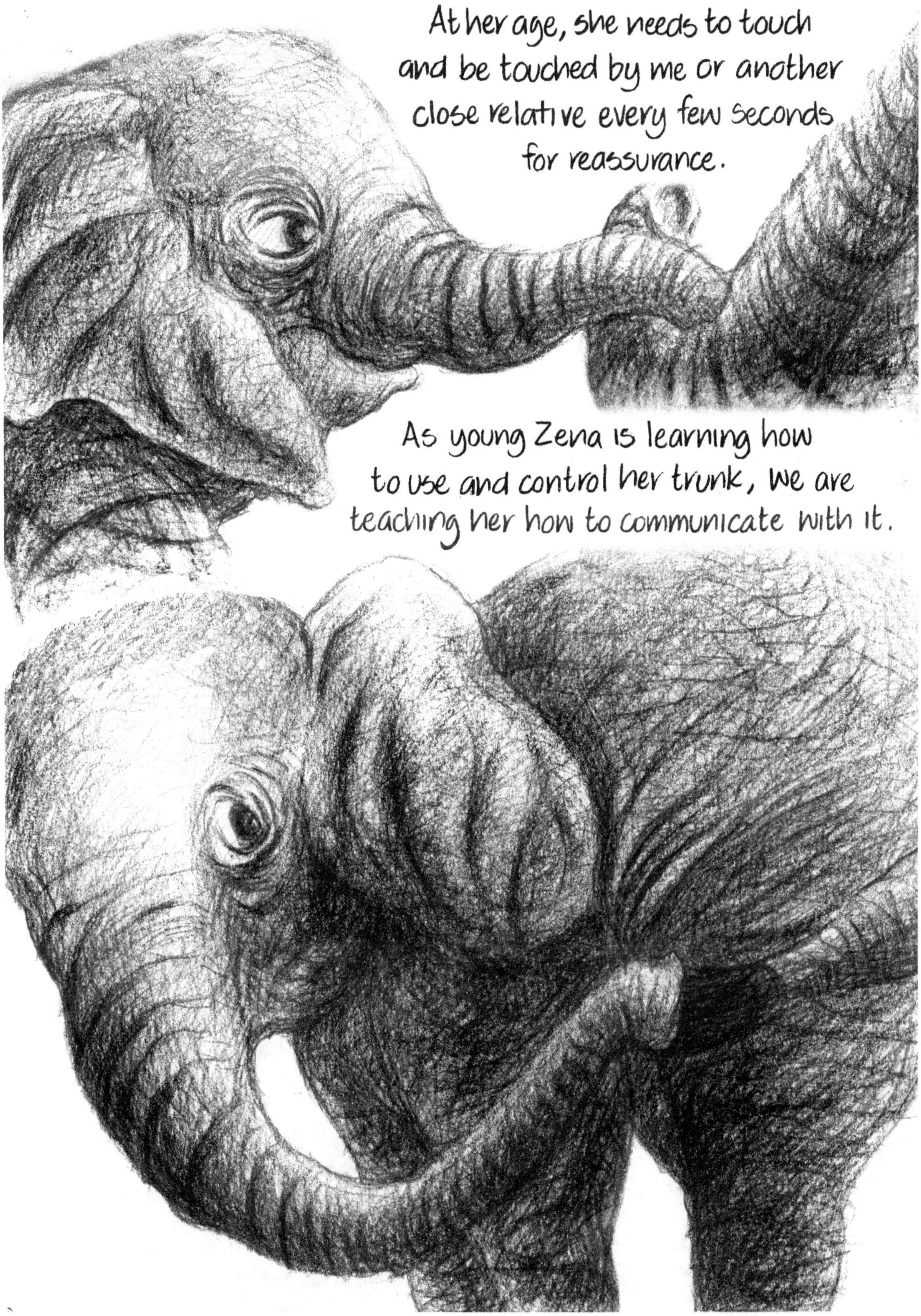
At her age, she needs to touch
and be touched by me or another
close relative every few seconds
for reassurance.
As young Zena is learning how
to use and control her trunk, we are
teaching her how to communicate with it.

Communication is an essential aspect of our survival. It is achieved by the efficient use of all our senses. Communicating is important because it enables us to keep track of one another, defending territories and alerting the group of danger. Although we can make a wide range of sounds, we mostly communicate through low-frequency sounds like growling and rumbling.

Our ears can hear sounds and,
in conjunction with the soles
of our feet and trunks,
sense vibrations
over long distances.

On average, we can hear
another elephant's call
at 2.5 miles away.

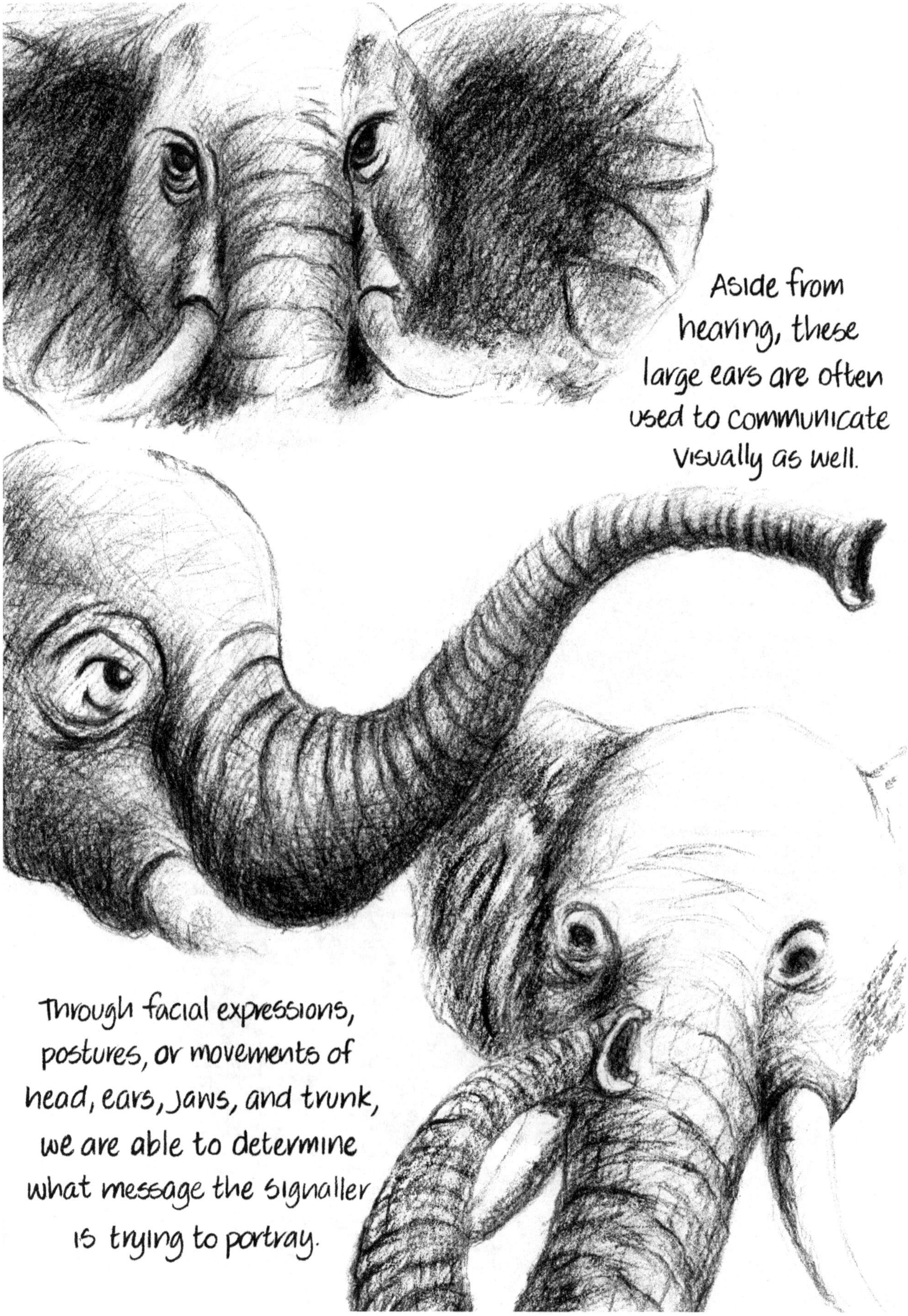

Aside from hearing, these large ears are often used to communicate visually as well.

Through facial expressions, postures, or movements of head, ears, jaws, and trunk, we are able to determine what message the signaller is trying to portray.

Not only do our senses provide the means for recognition and memory, but our family unit supplies the teaching blocks for young Zena. She learns by watching and mimicking our actions. These lessons will guide her to become a strong fearless leader.

SHASHWAT MISHRA

At Sea
Rafting: Oceanic dispersal via floating material
Algae, wood, volcanic pumice, and human debris all make fine rafts for animals to ride to far off lands.
By Alyssa Lee Suzumura

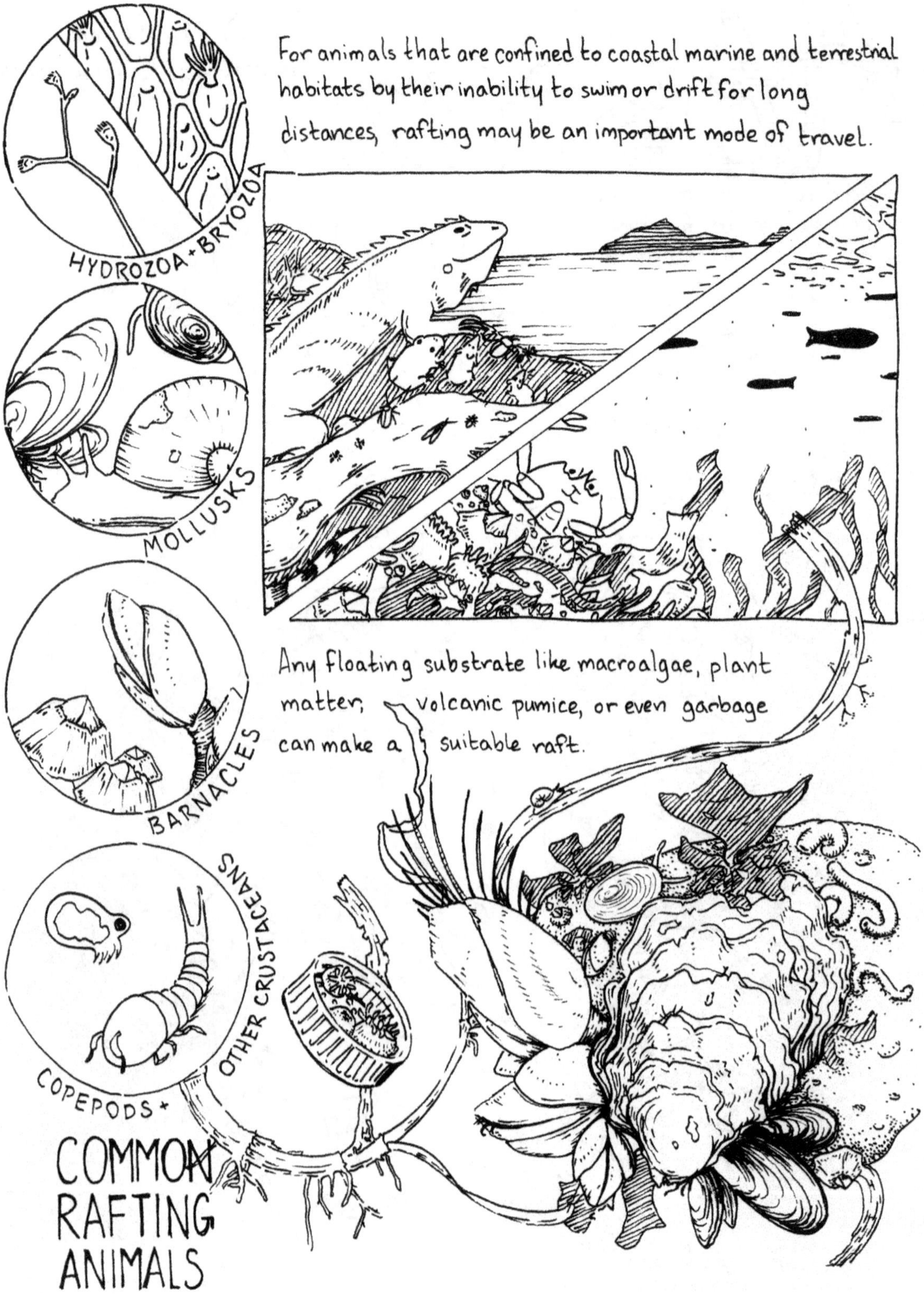
For animals that are confined to coastal marine and terrestrial habitats by their inability to swim or drift for long distances, rafting may be an important mode of travel.
HYDROZOA + BRYOZOA
MOLLUSKS
BARNACLES
COPEPODS + OTHER CRUSTACEANS
Any floating substrate like macroalgae, plant matter, volcanic pumice, or even garbage can make a suitable raft.
COMMON RAFTING ANIMALS

Though it's less common, terrestrial animals also raft!
Green iguanas arriving on the island of Anguilla from Guadaloupe following a 1995 hurricane is one of the only witnessed accounts of animals arriving by raft.

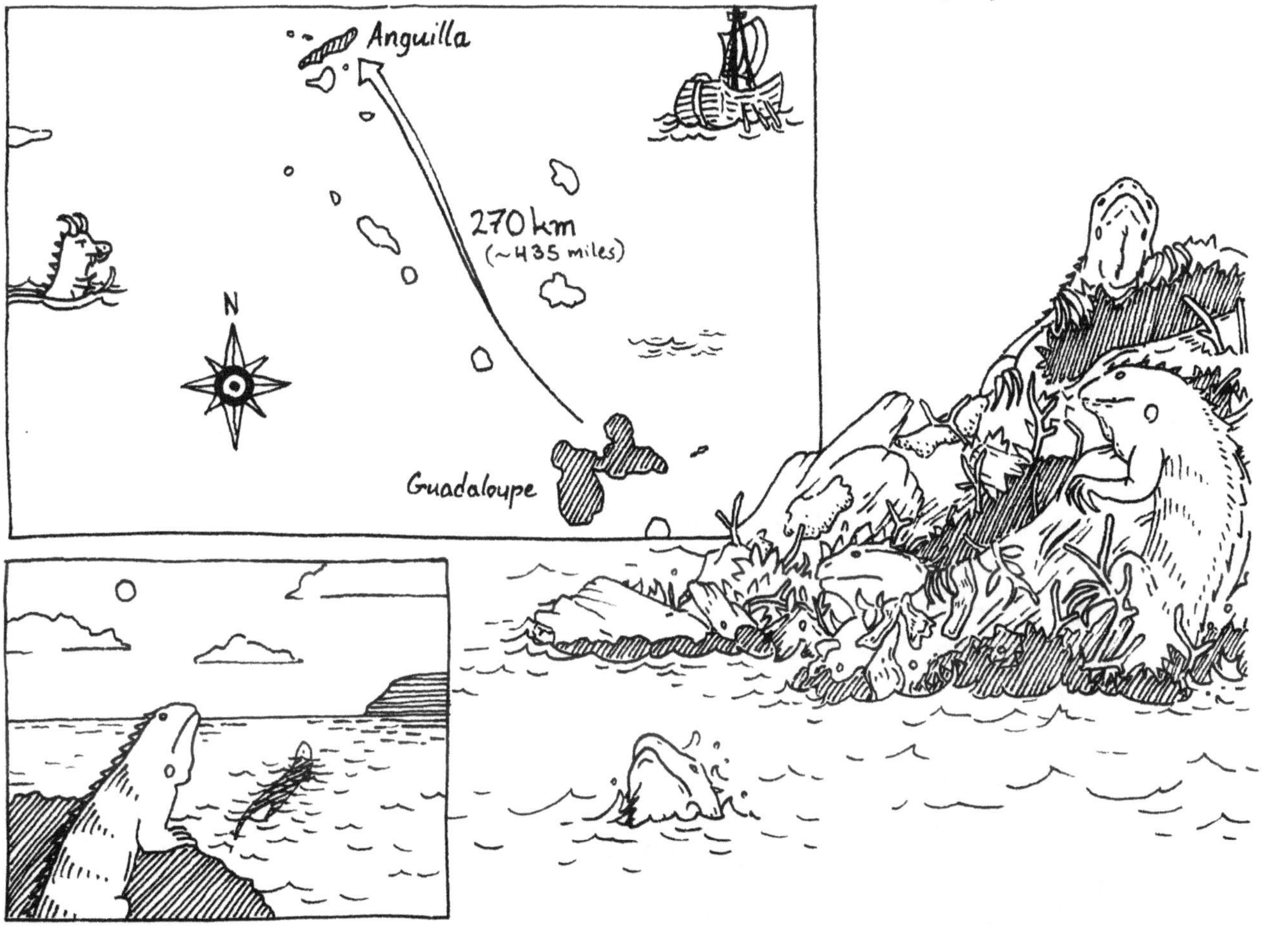

As many as 15 iguanas were reported on the island where they didn't previously exist, and they began reproducing within the following couple of years.

There is also testimony of wood-boring insects arriving on islands via floating wood, and recent genetic evidence has indicated that monkeys may have travelled from Africa to South America by rafting.

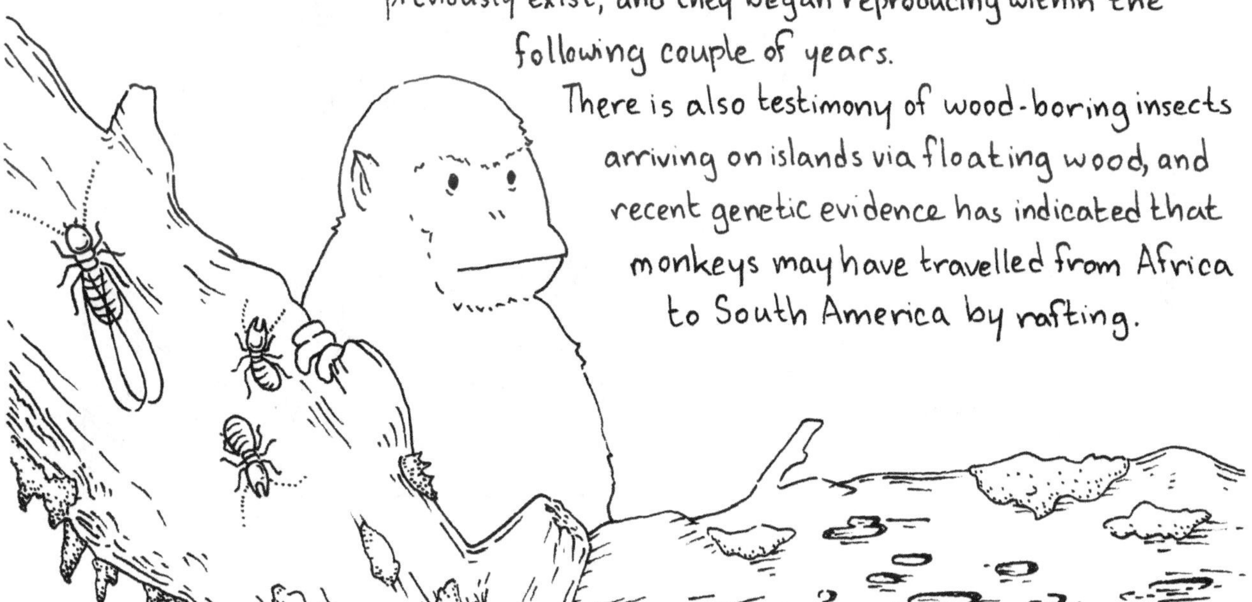

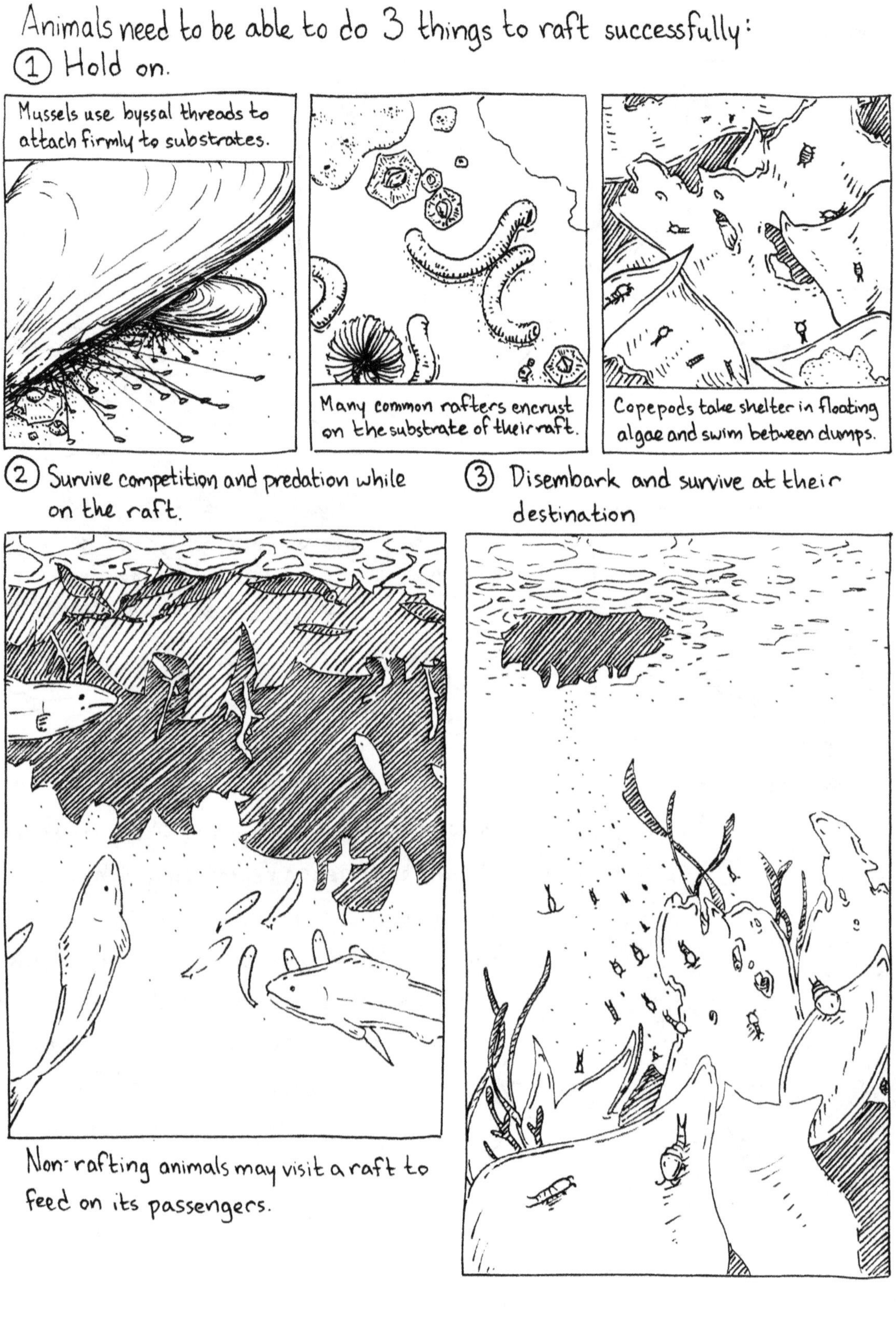
Animals need to be able to do 3 things to raft successfully:
① Hold on.
Mussels use byssal threads to attach firmly to substrates.
Many common rafters encrust on the substrate of their raft.
Copepods take shelter in floating algae and swim between clumps.
② Survive competition and predation while on the raft.
Non-rafting animals may visit a raft to feed on its passengers.
③ Disembark and survive at their destination

Life at sea can be difficult, and it seems that animals' chances of successful rafting are pretty low. Rafters face many challenges:

Despite these challenges, enough critters are able to survive that we have found rafting to be an important source of biodiversity and gene exchange in coastal marine and terrestrial habitats.

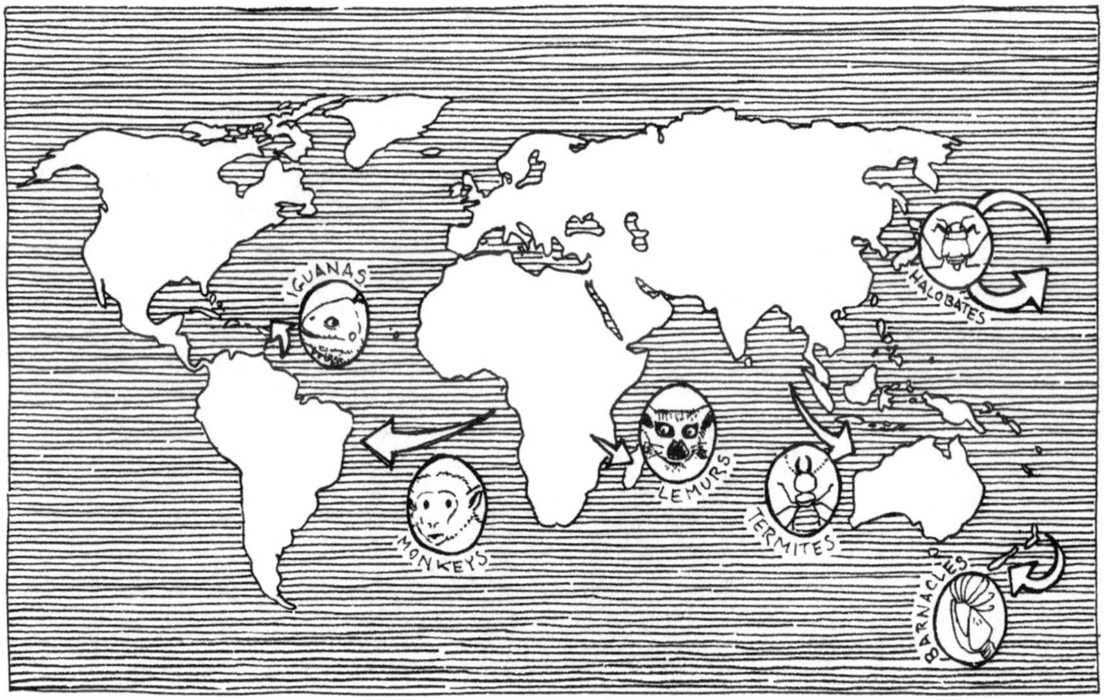

ARTHROPOD EXPLORERS

Rafting insects and crustaceans

What is Cryptozoology?

By Patricia Maldonado

From the Greek word *krypto* meaning "to hide," cryptozoology is the study of cryptids, hidden animals not accepted or recognized by mainstream science.

Creatures like Bigfoot or The Loch Ness Monster are popularly categorized as cryptids but actually belong in the category of paracryptozoology, which deals with beasts possessing otherworldly or supernatural abilities.

Cryptozoology has developed a stigma amongst scientists and zoologist because of eccentrics, hoaxes, and monster hunting shows in pop culture. It is currently viewed as a pseudoscience, about as reputable as astrology.

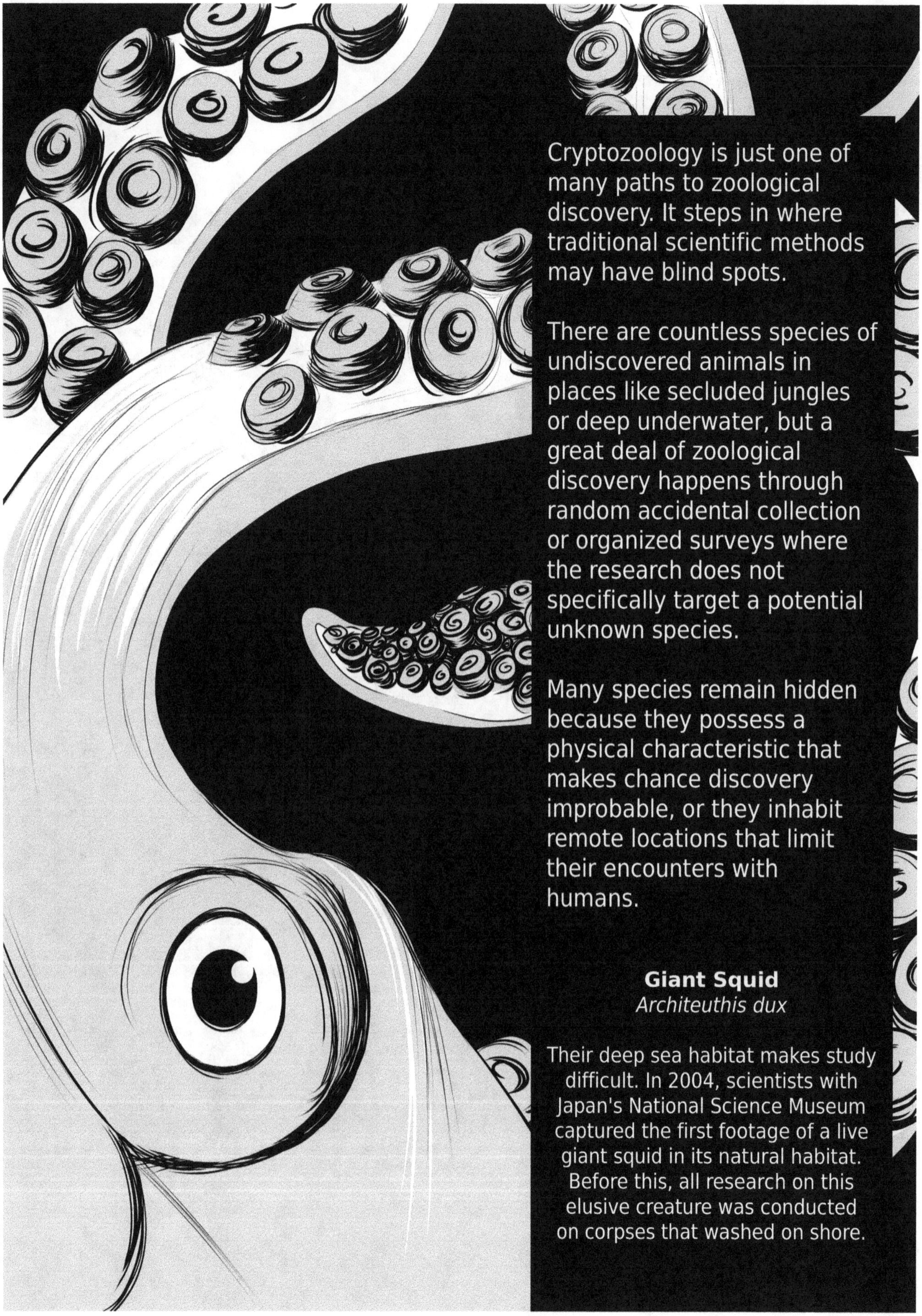

Cryptozoology is just one of many paths to zoological discovery. It steps in where traditional scientific methods may have blind spots.

There are countless species of undiscovered animals in places like secluded jungles or deep underwater, but a great deal of zoological discovery happens through random accidental collection or organized surveys where the research does not specifically target a potential unknown species.

Many species remain hidden because they possess a physical characteristic that makes chance discovery improbable, or they inhabit remote locations that limit their encounters with humans.

Giant Squid
Architeuthis dux

Their deep sea habitat makes study difficult. In 2004, scientists with Japan's National Science Museum captured the first footage of a live giant squid in its natural habitat. Before this, all research on this elusive creature was conducted on corpses that washed on shore.

Kim Qui, the Golden Turtle God of Vietnamese legend, is known today as the Hoàn Kiếm turtle *(Rafetus leloii)*.
The Egyptian god Set is most commonly represented by the Sha, an unknown animal depicted with large, triangular ears which resemble those of a fennec fox.
The Aztec Ahuizotl, a carnivorous aquatic monster, is speculated to have been a type of river otter.
Cryptozoology focuses on ethnoknown species: undiscovered animals that have had some prior contact with humans. These might be folkloric animals from a peoples' belief system or those mentioned in an explorer's journal.
Cryptozoologists analyze these reported sightings, determine which mystery animals might be real unknown biological species, and develop a methodology for obtaining concrete physical evidence of their existence.

Cryptids are divided into four main categories:

Animals similar to known species but with obvious phenotypic differences.

The marozi is believed to be distinct race of lion with a spotted coat adapted for forests rather than the savanna.

Animals similar to a known species but outside a normally accepted geographic range.

In 2006, 125,000 lowland gorillas were discovered in a swamp in the Northern Republic of Congo. Scientists acted on a tip by hunters and traveled 80 kilometers off the main road to find double the number of the endangered gorillas previously thought to survive worldwide.

Animals similar to a known, extinct species.

The thylacine, commonly known as the Tasmanian tiger, was hunted to extinction in the wild in 1930, with the last known member of the species dying in captivity in 1936. In the 80 years since the species became extinct, there have been thousands of unconfirmed sightings.

Animals that appear unrelated to any known or extinct species.

The okapi were first described in 1887, in the journal of explorer Henry Morton Stanley. Its unusual description caused many to doubt the animal's existence, until a specimen was sent to London in 1901.

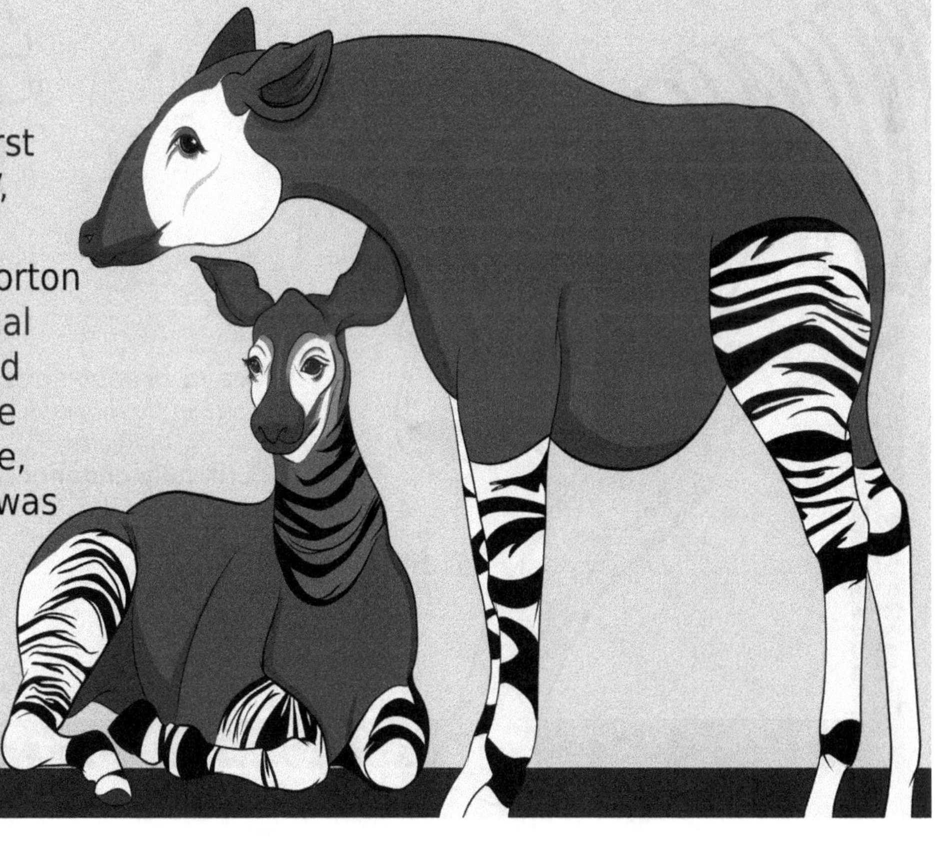

Cryptozoology has a strong focus on conservationism. Animal extinction happens naturally, and it is estimated that we lose 1 to 5 species a year. Deforestation or the introduction of invasive species to an environment has multiplied the natural rate of extinction. It is estimated we are losing dozens of species a day due to human actions.

Countless species go extinct before they are discovered, and many cryptozoologists hope that inspiring interest in the discovery of new animals will also inspire a greater interest in conservation efforts.

In order to protect both undiscovered and endangered animals, we must first protect our environment.

Pyrenean ibex
(Capra pyrenaica pyrenaica)
Extinct

Macaya breast-spot frog
(Eleutherodactylus thorectes)
Critically endangered

Javan tiger
(Panthera tigris sondaica)
Extinct

THE FANTASTIC ANNUAL MIGRATIONS of NORTH AMERICAN
SHOREBIRDS
Every year, shorebirds fly arduous round-trip migrations, hailing from as far as the southernmost tip of South America to the cold tundra above the Arctic Circle near the North Pole.
written and illustrated by Anita K. Boyle
Shorebirds are wading birds who poke their bills into the surf and beaches to find their food.
They nest in shallow holes of sand gravel, leaves, or grass.
More than half of all shorebirds on the Western Hemisphere migrate from Central or South America to the Arctic. These are very tough birds.
The 50 shorebird species who call North America home come from five families of the order Charadriiformes, or "waders."
Meet six of them in the next few pages.

LEAST SANDPIPER
Calidris minutilla

This smallest of shorebirds, the least sandpiper, has the nickname "Peeps" because of its diminutive size.

Size & Shape: Up to 4 inches (10 cm) short, stalky with a hunched posture
Annual Round-Trip Migration: 16,500 miles (26,550 km)
Breeding: Nests between late May and late August across northernmost North America in sub-arctic tundra and boreal forests
Winters: Coastal and interior from Washington to Virginia south to Chile and Brazil
Status: Least concern, though its population is showing a decline

Once eggs are laid in the nest, the male is the primary incubator. The chicks can fly within 15 days after hatching.

This little sandpiper moseys along the shore while chowing down on invertebrates such as amphipods, gastropods, and flies.

GREATER YELLOWLEGS

Tringa melanoleuca

Both the male and the female share incubation duties as well as raising the chicks, who will be ready to fly within 20 days.

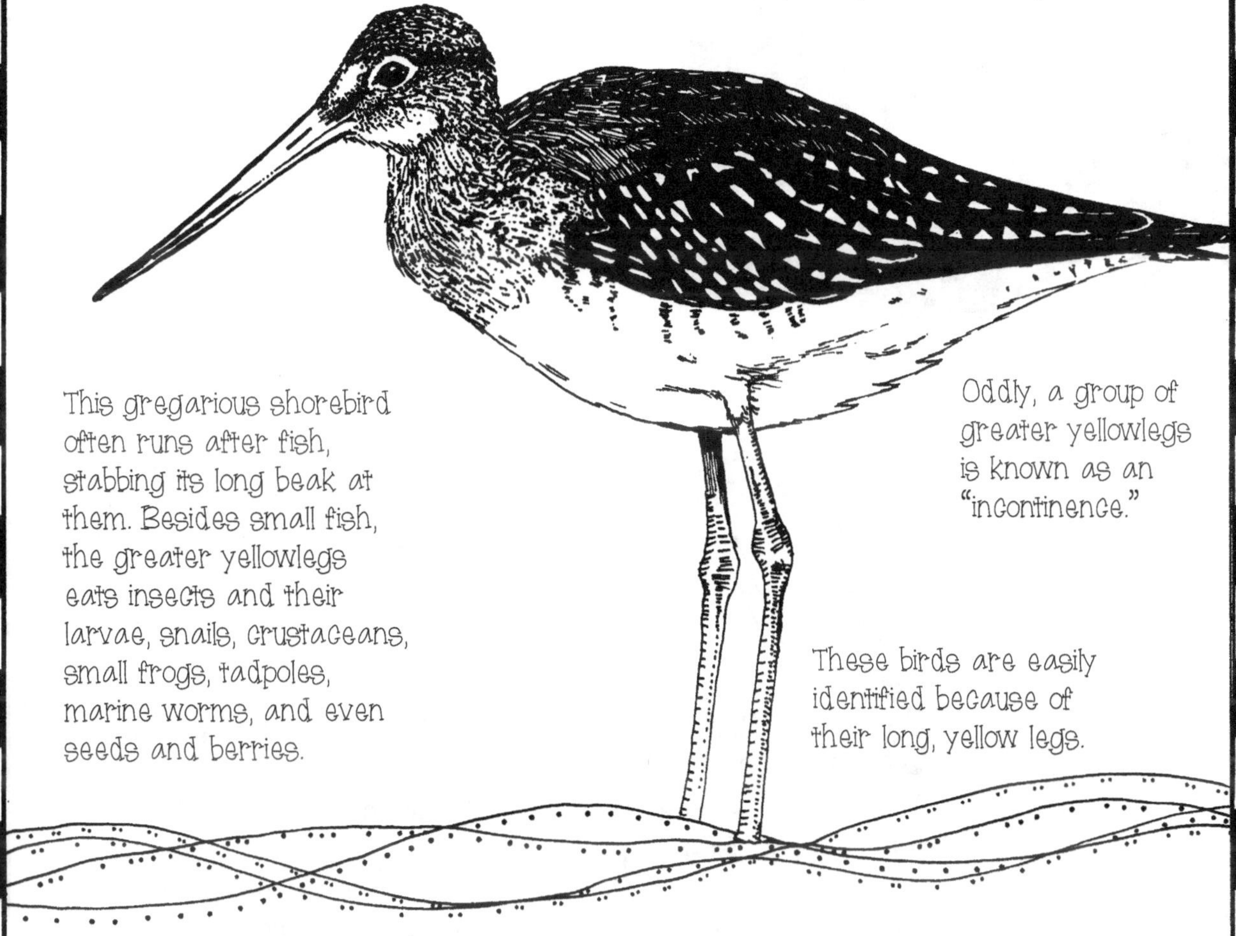

This gregarious shorebird often runs after fish, stabbing its long beak at them. Besides small fish, the greater yellowlegs eats insects and their larvae, snails, crustaceans, small frogs, tadpoles, marine worms, and even seeds and berries.

Oddly, a group of greater yellowlegs is known as an "incontinence."

These birds are easily identified because of their long, yellow legs.

Size & Shape: 18 inches (46 cm) tall, with a long beak and bright yellow legs
Annual Round-Trip Migration: 16,000 miles (25,750 km)
Breeding: South Alaska east to Newfoundland in boreal forests (bogs, meadows, muskegs, and sometimes subarctic tundra)
Winters: South British Columbia east to Connecticut and south to Central and South American and the West Indies
Status: Least concern

AMERICAN AVOCET

Recurvirostra americana

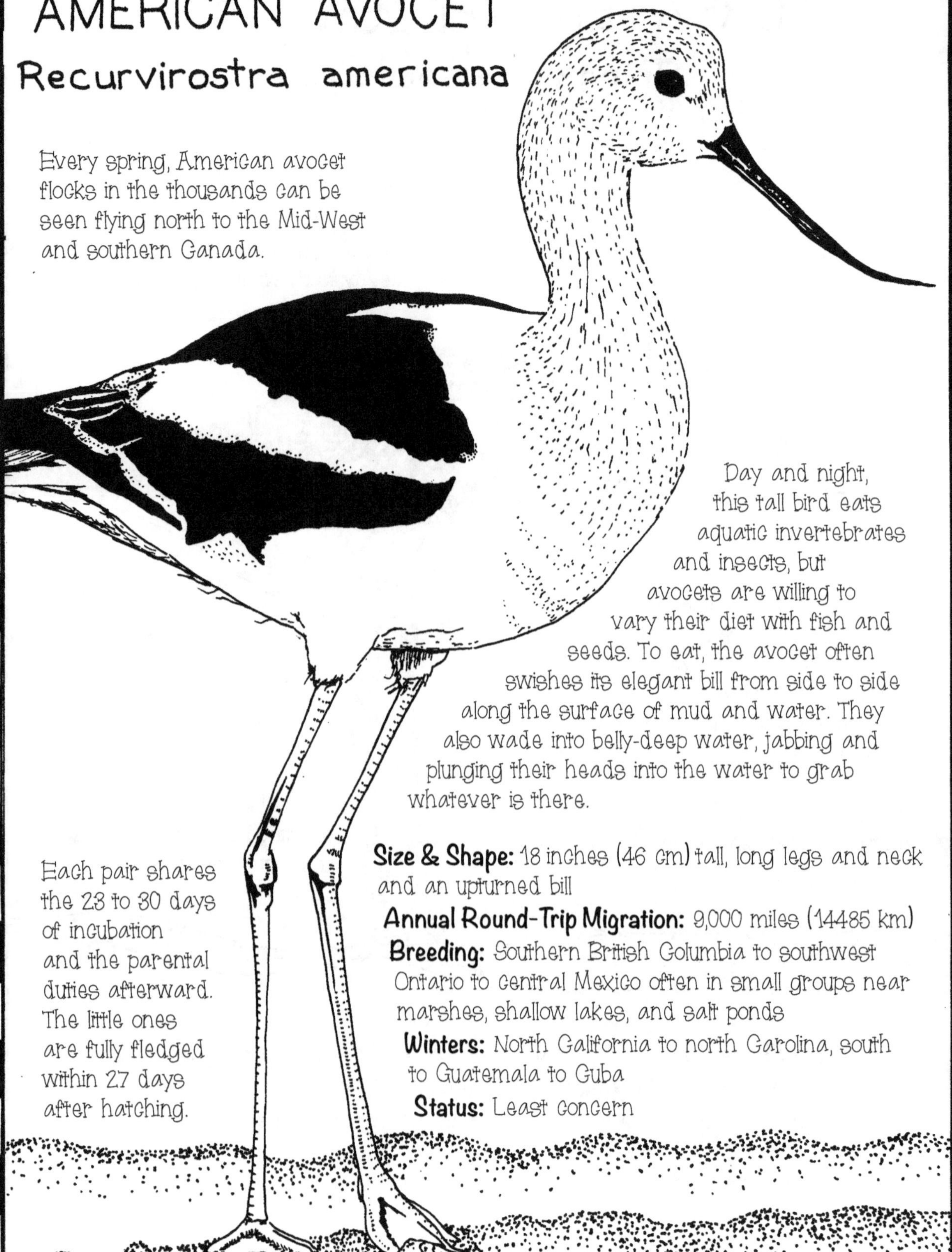

Every spring, American avocet flocks in the thousands can be seen flying north to the Mid-West and southern Canada.

Day and night, this tall bird eats aquatic invertebrates and insects, but avocets are willing to vary their diet with fish and seeds. To eat, the avocet often swishes its elegant bill from side to side along the surface of mud and water. They also wade into belly-deep water, jabbing and plunging their heads into the water to grab whatever is there.

Each pair shares the 23 to 30 days of incubation and the parental duties afterward. The little ones are fully fledged within 27 days after hatching.

Size & Shape: 18 inches (46 cm) tall, long legs and neck and an upturned bill

Annual Round-Trip Migration: 9,000 miles (14485 km)

Breeding: Southern British Columbia to southwest Ontario to central Mexico often in small groups near marshes, shallow lakes, and salt ponds

Winters: North California to north Carolina, south to Guatemala to Cuba

Status: Least concern

WHIMBREL

Numenius phaeopus

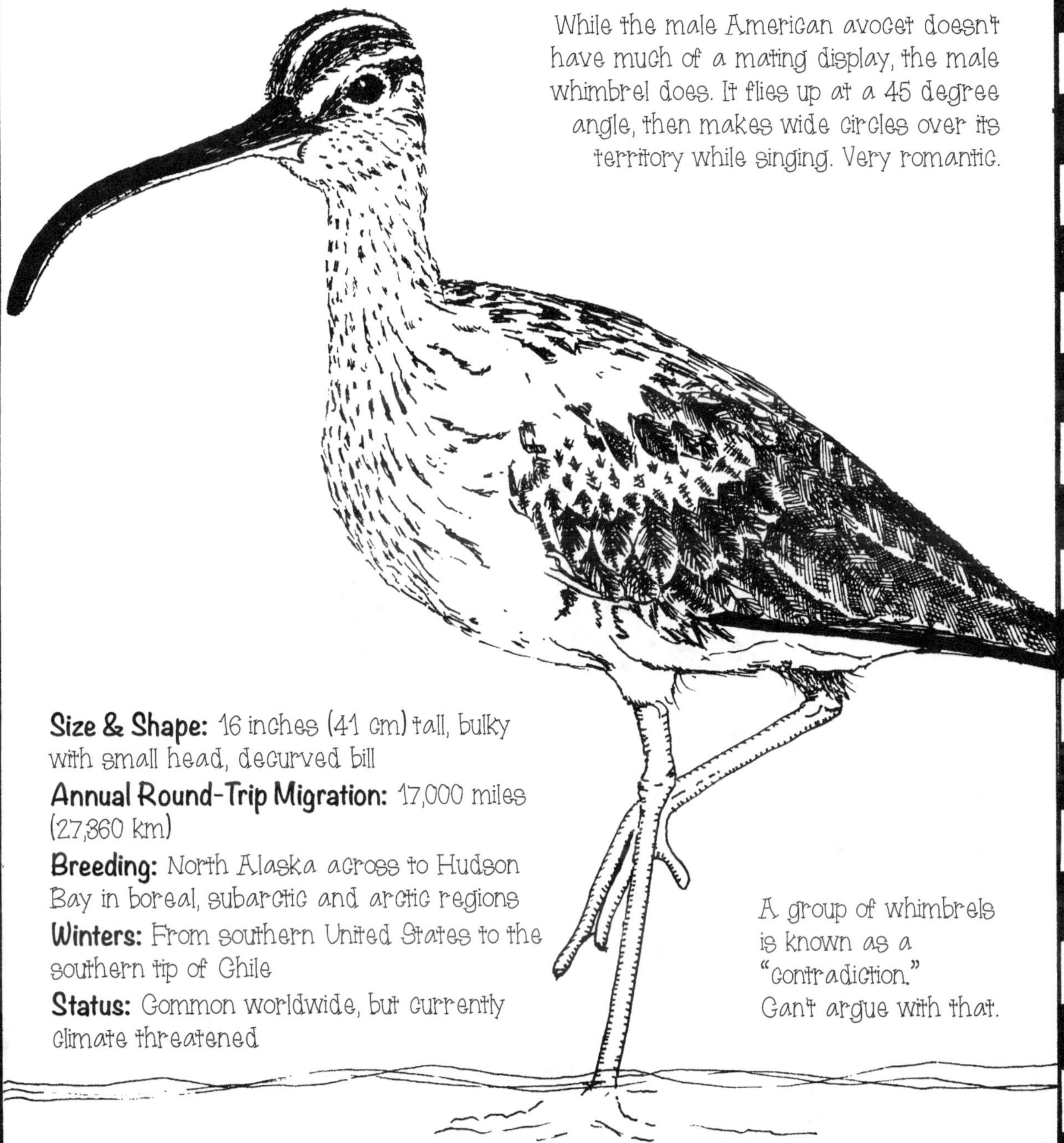

While the male American avocet doesn't have much of a mating display, the male whimbrel does. It flies up at a 45 degree angle, then makes wide circles over its territory while singing. Very romantic.

Size & Shape: 16 inches (41 cm) tall, bulky with small head, decurved bill
Annual Round-Trip Migration: 17,000 miles (27,360 km)
Breeding: North Alaska across to Hudson Bay in boreal, subarctic and arctic regions
Winters: From southern United States to the southern tip of Chile
Status: Common worldwide, but currently climate threatened

A group of whimbrels is known as a "contradiction." Can't argue with that.

A steady diet of crabs, worms, mussels, and marine invertebrates will keep the whimbrel on its toes. Sometimes, one will cram its bill all the way into a fiddler crab burrow to get a tasty treat.

RUDDY TURNSTONE
Arenaria interpres

Size & Shape: 10 inches (25 cm) tall, chunky and colorful, with a short stout bill
Annual Round-Trip Migration: 17,000 miles (27,360 km)
Breeding: Rocky coasts and tundra of the Arctic tundra
Winters: Washington state to Newfoundland south to central Chile and Argentina
Status: Common on sandy and gravelly seacoasts

A ruddy turnstone will speed walk along shorelines, flipping over rocks and pebbles, to find a delicious meal of marine invertebrates, insects, carrion, eggs, garbage, plants, crustaceans, mollusks, snails, and almost anything that wiggles.

Males make "fake" nests on the ground to attract an eager mate to his territory, but the female lays eggs in a nest she makes herself. Once the eggs are in the nest, she does most of the 22 days of incubation, and he does most of the rearing of the brood.

The ruddy turnstone is one of the world's most widespread birds with a presence on every continent in the world except Antarctica.

RED KNOT
Calidris canutus

Size & Shape: 8 inches (20 cm) chunky with a blunt, short bill
Annual Round-Trip Migration: 19,000 miles (30,580 km)
Breeding: Arctic tundra on dry rocky ground with lots of low vegetation
Winters: From southern USA coasts through southern Argentina
Status: Once common, but recently declining rapidly due to environmental pressures. The United States Shorebird Conservation Program currently lists the red knot as "Highly Imperiled," and the International Union for Conservation of Nature put it on their "Red List" as "Near Threatened."

This bird-on-the-move eats mainly insects and vegetation on the tundra during the spring breeding period, but while migrating or wintering, it enjoys bivalves, small snails, worms, and crustaceans along the coasts.

The round-trip flight of the red knot is one of the longest annual migrations of any animal in the world.

In June or July, a pair of red knots shares incubation duties for 21 days. Then the male takes over most the brood-rearing duties.

During the spring migration north, the most important stopover for these long-distance migrators is Delaware Bay. Red knots arrive thin and hungry, and need to double their weight quickly. They gorge on energy-rich horseshoe crab eggs and then head up to their Arctic breeding grounds. But horseshoe crabs have been declining, mainly because of shoreline real estate development and commercial fishing.

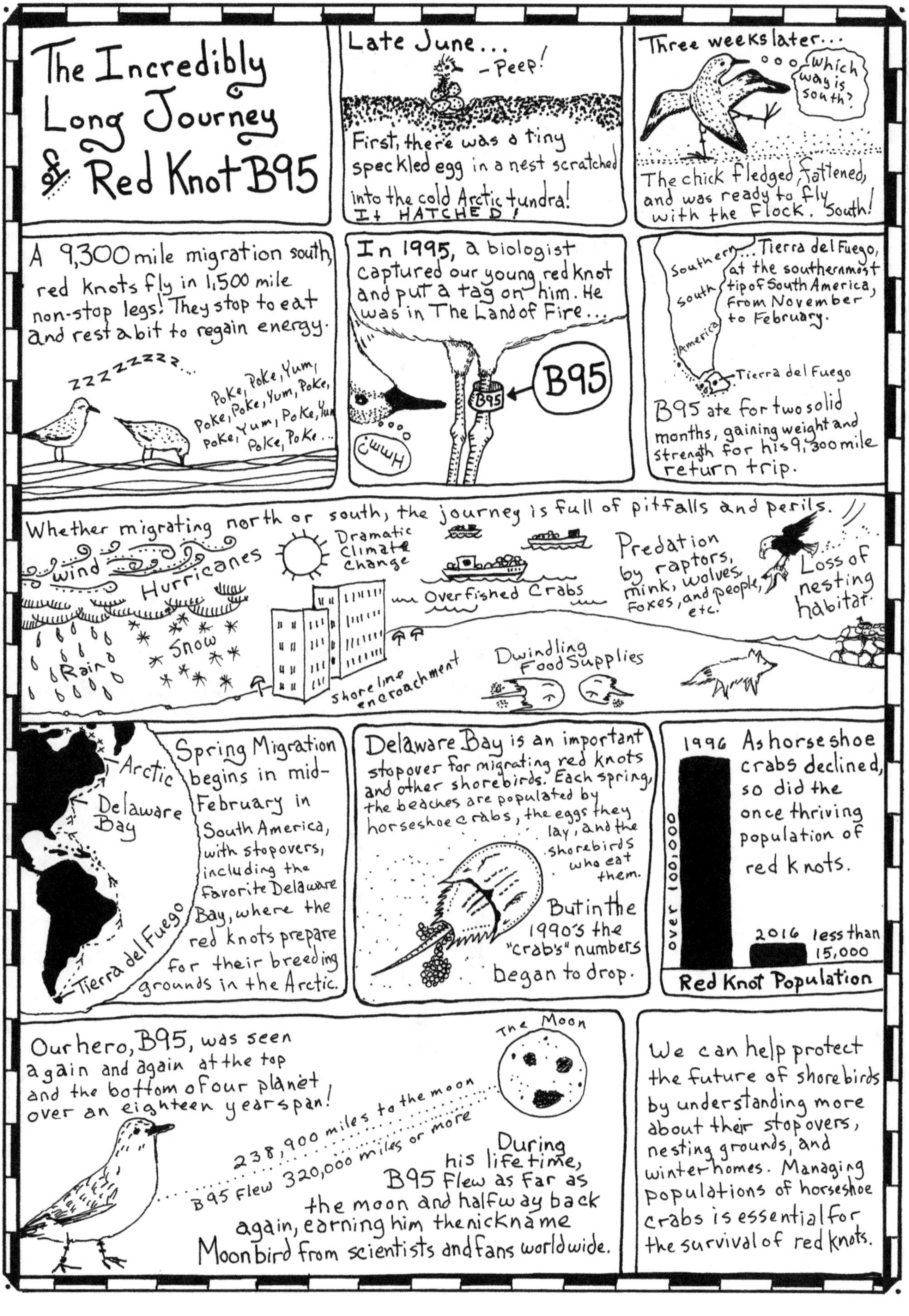
The Incredibly Long Journey of Red Knot B95
Late June...
–Peep!
First, there was a tiny speckled egg in a nest scratched into the cold Arctic tundra! It HATCHED!
Three weeks later...
Which way is south?
The chick fledged, fattened, and was ready to fly with the flock. South!
A 9,300 mile migration south, red knots fly in 1,500 mile non-stop legs! They stop to eat and rest a bit to regain energy.
ZZZZZZZZZ...
Poke, Poke, Yum, Poke, Poke, Yum, Poke, Poke, Yum, Poke, Yum Poke, Poke, Poke...
In 1995, a biologist captured our young red knot and put a tag on him. He was in The Land of Fire...
B95
B95
Hmm?
Southern South America
...Tierra del Fuego, at the southernmost tip of South America, from November to February.
Tierra del Fuego
B95 ate for two solid months, gaining weight and strength for his 9,300 mile return trip.
Whether migrating north or south, the journey is full of pitfalls and perils.
Wind
Hurricanes
Dramatic climate change
Overfished Crabs
Predation by raptors, mink, wolves, foxes, and people, etc.
Loss of nesting habitat
Rain
Snow
Shoreline encroachment
Dwindling Food Supplies
Arctic
Delaware Bay
Tierra del Fuego
Spring Migration begins in mid-February in South America, with stopovers, including the favorite Delaware Bay, where the red knots prepare for their breeding grounds in the Arctic.
Delaware Bay is an important stopover for migrating red knots and other shorebirds. Each spring, the beaches are populated by horseshoe crabs, the eggs they lay, and the shorebirds who eat them.
But in the 1990's the "crab's" numbers began to drop.
1996
over 100,000
2016 less than 15,000
Red Knot Population
As horseshoe crabs declined, so did the once thriving population of red knots.
Our hero, B95, was seen again and again at the top and the bottom of our planet over an eighteen year span!
The Moon
238,900 miles to the moon
B95 flew 320,000 miles or more
During his lifetime, B95 flew as far as the moon and halfway back again, earning him the nickname Moonbird from scientists and fans worldwide.
We can help protect the future of shorebirds by understanding more about their stopovers, nesting grounds, and winter homes. Managing populations of horseshoe crabs is essential for the survival of red knots.

The Day I Saw the Mountain Beaver
by N. DUPILLE 2016
I WAS HIKING WITH A GROUP OF CAMP COUNSELORS ONE JULY.
thimbleberry leaves make excellent TP
I STOPPED TO PEE IN A LITTLE CLEARING WITH A NICE VIEW OF THE PILCHUCK RIVER.
fairies?!
I WAS BLISSFULLY MID-STREAM WHEN I SAW A PAIR OF TINY HANDS EMERGE FROM A PATCH OF RUSTLING LEAVES.
RUSTLE
RUSTLE
!
THE HANDS WERE CONNECTED TO A SMALL CREATURE MUNCHING AWAY AT SOME GREENERY.

what IS this tiny adorable animal??
I WANTED TO GET UP OR SHOUT FOR SOMEONE SO I COULD ASK WHAT IT WAS, BUT GIVEN MY POSITION, ALL I COULD DO WAS SIT AND BE CAPTIVATED BY THIS LITTLE CRITTER.
WHEN I GOT BACK TO CAMP, I ASKED THE DIRECTOR ABOUT IT.
it was a mountain beaver!
is that even a real animal?!
IT TURNED OUT HE WASN'T MESSING WITH ME.
closely related to squirrel
hi.
not at all related to beaver
THE MOUNTAIN BEAVER IS REAL, ALIVE, AND THRIVING IN THE PACIFIC NORTHWEST.

a long time ago...
T-Rex
mountain beaver
(unconfirmed)
OUR SMALL FURRY FRIEND IS ONE OF THE OLDEST LIVING SPECIES OF RODENT.
POTENTIAL MOUNTAIN BEAVER SIGHTINGS:
burrow
25 ft
THEY LIVE UNDERGROUND AND FORAGE CLOSELY AROUND THEIR BURROWS FOR FOOD AND WATER.
it's all yours!
the burrow
THEY PLAY A KEY ROLE IN NATURE: MOVING SOIL, FERTILIZING IT, AND LEAVING THEIR BURROWS FOR OTHERS TO USE.
Natalie 2016
end.
I GOT LUCKY— SIGHTINGS ARE RARE. BUT THE NEXT TIME YOU'RE IN THE WOODS, LOOK FOR SOME SLENDER-TOED TRACKS. YOU JUST MIGHT GET LUCKY, TOO.

KAMI KOYAMATSU

DOCTOR, MY APHID'S NOT WELL. I WEBMD'D IT, AND I'M PRETTY SURE IT'S HEPATIC VENO-OCCLUSIVE DISEASE.
I THINK MY WHOLE HERD'S GOT IT.
COOCHIE COOCHIE COO.
COO!
IT'S NOT HEPATIC VENO-OCCLUSIVE DISEASE. APHIDS DON'T GET THAT.
BUT THEY DO GET PREGNANT, AND THAT'S EXACTLY WHAT THIS LITTLE GIRL IS.
AREN'T YOU? YOU LITTLE CUTIE?
SQUEE!
CAITLIN HOFMEISTER & LAUREN NORBY
present...

Babies Having Babies

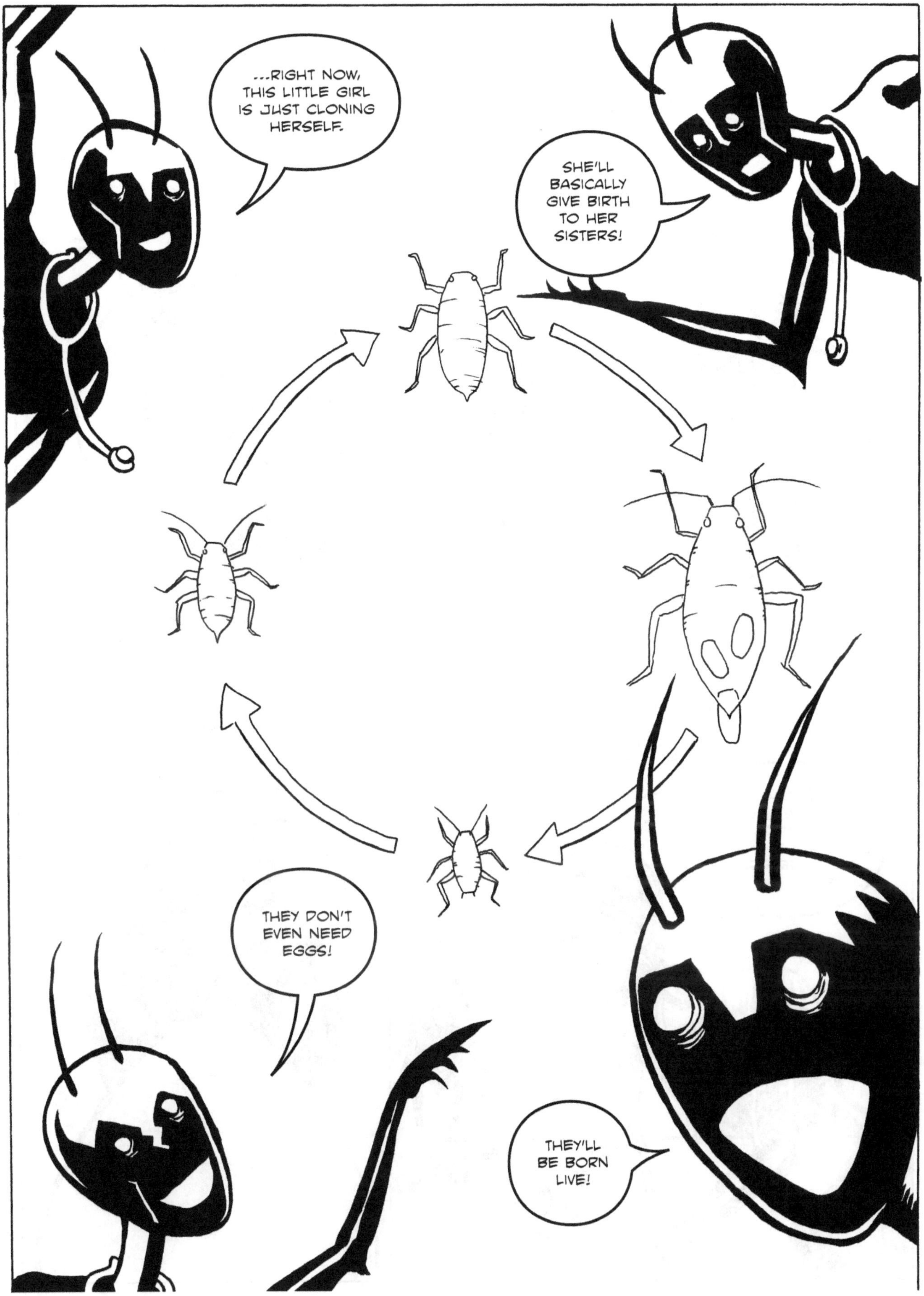
...RIGHT NOW, THIS LITTLE GIRL IS JUST CLONING HERSELF.
SHE'LL BASICALLY GIVE BIRTH TO HER SISTERS!
THEY DON'T EVEN NEED EGGS!
THEY'LL BE BORN LIVE!

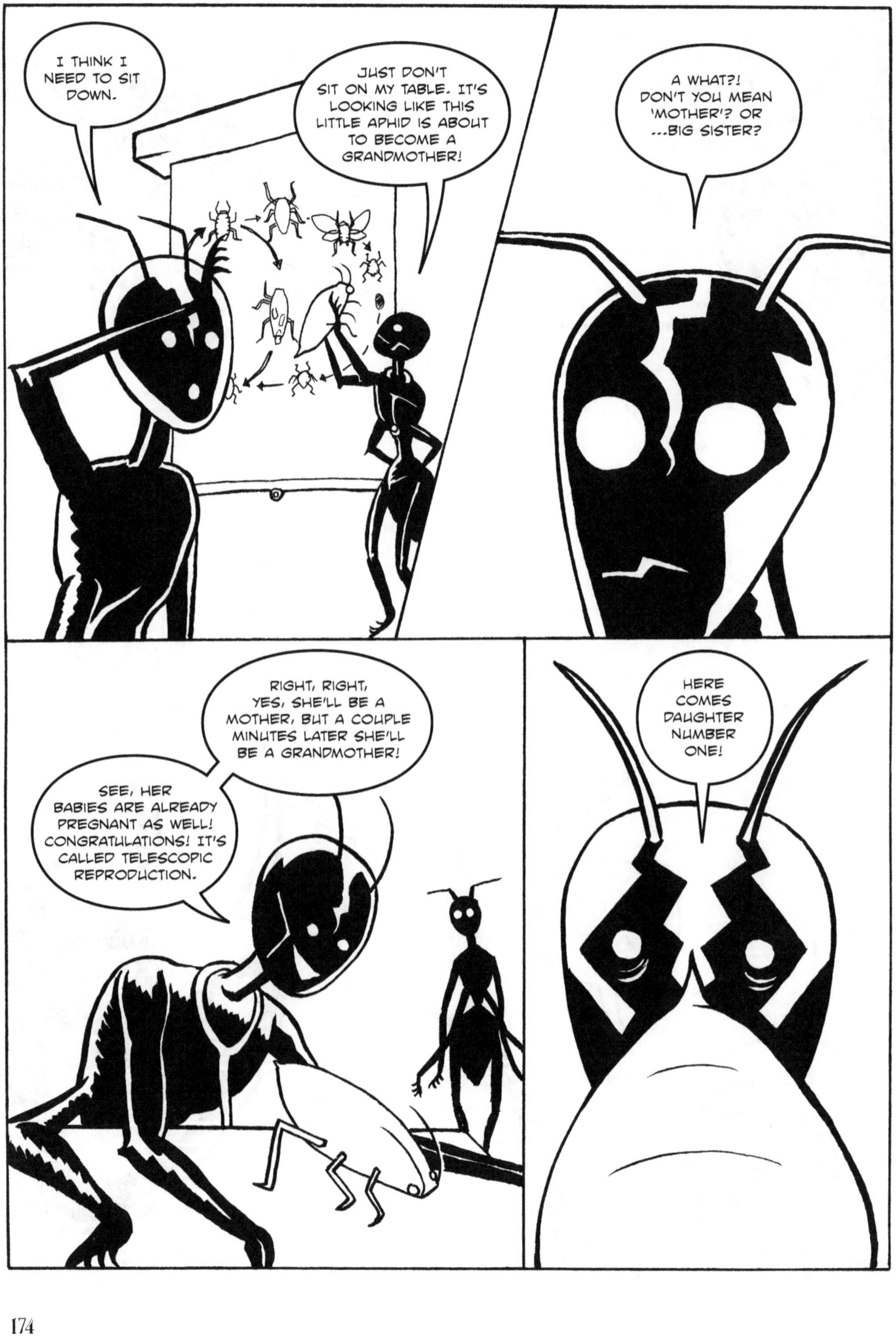

I THINK I NEED TO SIT DOWN.
JUST DON'T SIT ON MY TABLE. IT'S LOOKING LIKE THIS LITTLE APHID IS ABOUT TO BECOME A GRANDMOTHER!
A WHAT?! DON'T YOU MEAN 'MOTHER'? OR ...BIG SISTER?
RIGHT, RIGHT, YES, SHE'LL BE A MOTHER, BUT A COUPLE MINUTES LATER SHE'LL BE A GRANDMOTHER!
SEE, HER BABIES ARE ALREADY PREGNANT AS WELL! CONGRATULATIONS! IT'S CALLED TELESCOPIC REPRODUCTION.
HERE COMES DAUGHTER NUMBER ONE!

DOCTOR! WILL THIS EVER STOP?
I THINK THIS MIGHT BE THE LAST ONE!

THIS ONE'S NOT PREGNANT?
HOW CAN YOU TELL SHE'S THE LAST?
SHE HAS WINGS.
OFF TO REPRODUCE SEXUALLY THIS TIME. LUCKY LITTLE GIRL. LAY YOUR EGGS FAR AND WIDE.
SO, SEE YOU NEXT SUMMER?

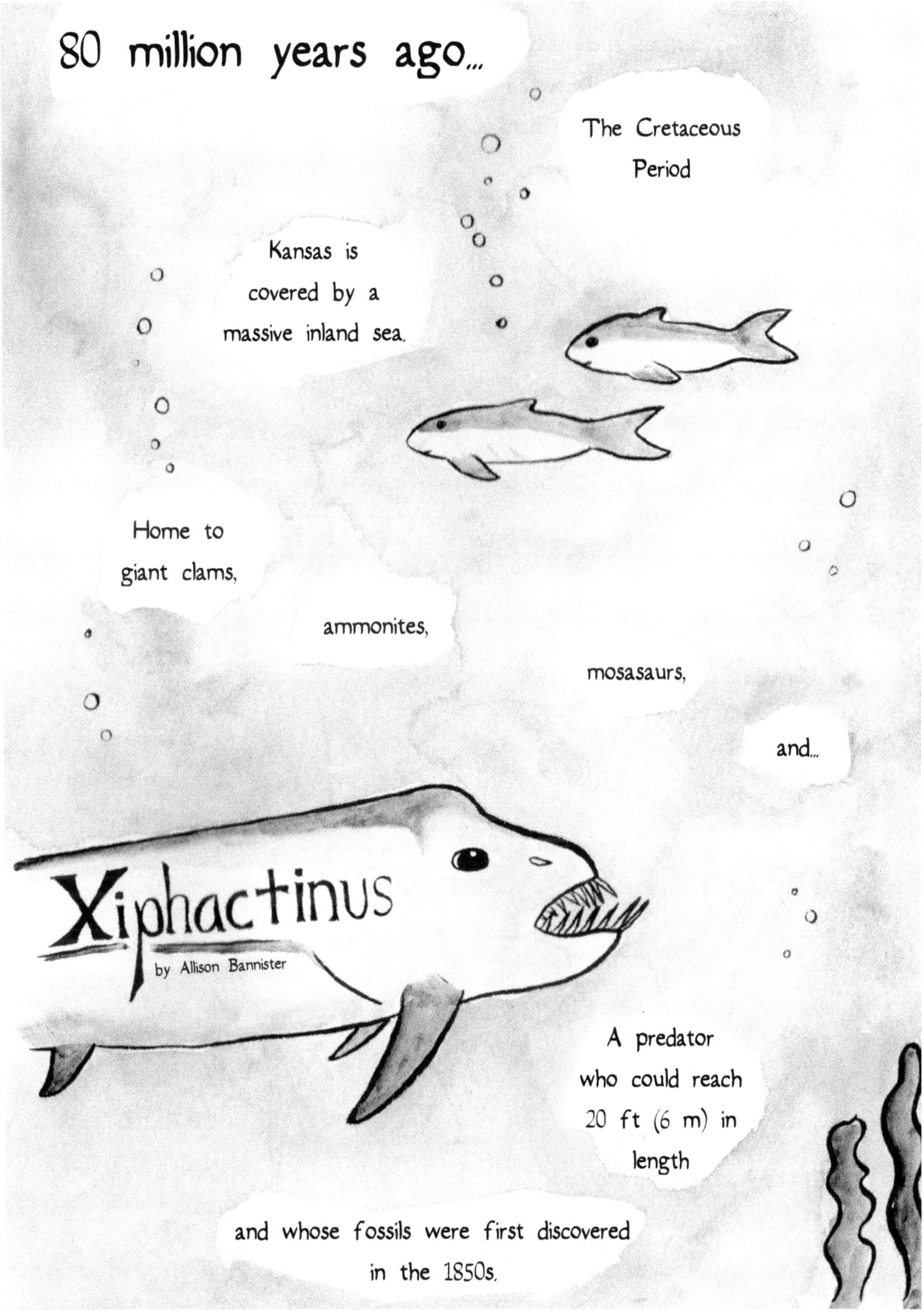
80 million years ago...
The Cretaceous Period
Kansas is covered by a massive inland sea.
Home to giant clams,
ammonites,
mosasaurs,
and...
Xiphactinus
by Allison Bannister
A predator who could reach 20 ft (6 m) in length
and whose fossils were first discovered in the 1850s.

Of all Xiphactinus specimens, the most remarkable was discovered in June of 1952, in Gove County Kansas, by George Sternberg.
The 14 ft predator had hunted down a 6 ft Gillicus,
swallowed the smaller fish whole,
and promptly died.

80 million years later, it was carefully uncovered,

encased in plaster,

and removed from the limestone bed in two pieces.

From there it was moved by truck to join Sternberg's fossil collection on display at Fort Hays State University.

In 1999, it was moved to it's current home in the Sternberg Museum of Natural History.

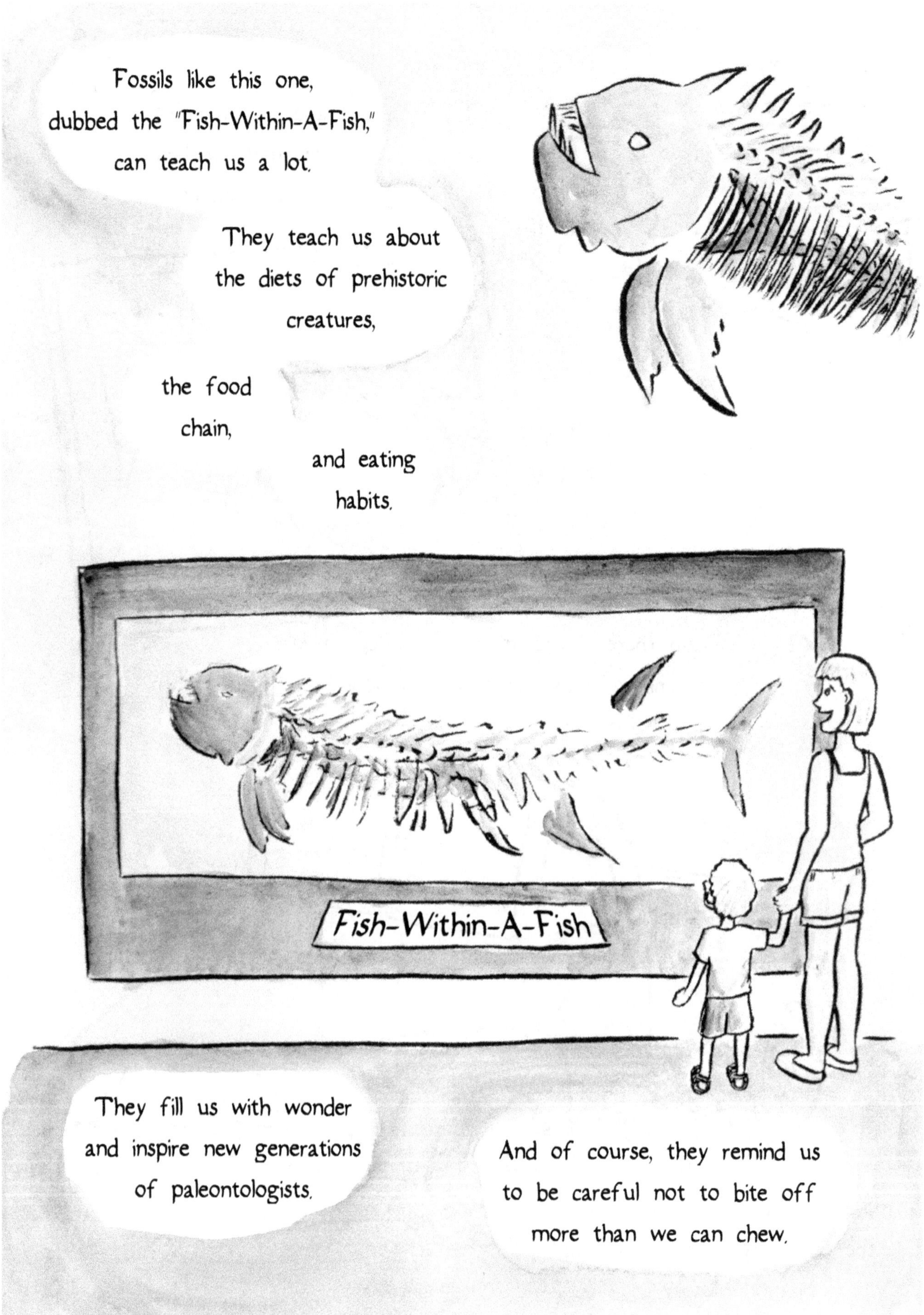
Fossils like this one, dubbed the "Fish-Within-A-Fish," can teach us a lot.
They teach us about the diets of prehistoric creatures,
the food chain,
and eating habits.
Fish-Within-A-Fish
They fill us with wonder and inspire new generations of paleontologists.
And of course, they remind us to be careful not to bite off more than we can chew.

Cordyceps
- a comic about parasitic fungi -
by Simina Popescu
Ophiocordyceps unilateralis is a parasitic fungus that infects ants. It has been known to wipe out entire colonies.
Ants have evolved the ability to detect infection among their own.
Healthy ants will carry the dying one far away from the colony.
These infected hosts are hijacked by the fungus. Cordyceps controls the ant, forcing it into very specific behaviour.
It seeks out a humid and warm environment.
When the conditions are right (over 25°C/77°F, over 77% humidity, close to the tropical forest floor), the Cordyceps makes the insect climb.

In each case, the ant stops around 25 centimeters (9.8 inches) above the ground.
Still driven by the parasite, the ant fixes itself to a major vein on the underside of a leaf with its mandibles.
This death grip is never released.
In every case, from a specific point at the back of the neck, the fungus grows a branch, or **hyphae**.
In a few days, it produces a "fruiting" body, or a spore pod.
The Cordyceps releases spores, creating an infectious killing field below.
Any ants underneath the shower of spores are infected.

Each species of Cordyceps affects only a single species of ant.
microscopic image of Cordyceps spores (100 μm)
Cordyceps spores like these must have closely coevolved with the host species to be able to have such strong control over something as complex as a behavioural system.
The two species are estimated to have been evolving along-side each other for over 48 million years.
Ophiocordyceps unilateralis is prevented from devastating entire colonies by suffering from a secondary fungus, or hyperparasite, that feeds off and kills the Cordyceps itself.

There are approximately
400 species of Cordyceps.
Fossil evidence shows that this fungus's
infection of their choice host has been
occurring for at least 48 million years.
The more populous a species
becomes,the more likely it is to be
preyed on by a Cordyceps fungus.

The Degu

by Bridget Comeau

What is a Degu?
Degus [DAY-GOOs] are rodents that live in Chile...
Chile
... and are cousins to the guinea pig, chinchilla, and porcupine.

Degu Life

Degus are herbivorous (eating only plants)

Due to an abnormal insulin function, they can only digest sugars that are specific to plants.

Degus love eating flowers, like dandelions, marigolds, and rose petals.

Degus are highly social mammals, living in groups of five or more. They live in large burrow systems and are diurnal (active during the day)

Physiology

Degus grow to about the size of an adult human's palm.

They have brown fur, pale fur on their belly and light rings of fur around their eyes. Their tails are medium length, with short wiry hair that becomes longer at the tip.

Like their relatives the chinchilla, degus have orange enamel on their teeth. Researchers believe the enamel turns orange due to how their bodies process the plants they eat or their iron intake.

Physiology Cont.

Degu brains are nearly the same as those of their cousins, but the olfactory bulb (used to distinguish smells) is particularly developed.

Degu fur is camouflaged; however, their belly fur is lighter and UV reflective. They stand on their hind legs and expose their belly to signal for alarm calls.

History and Research
The degu was classified as a species in 1782 by Father Juan Ignacio Molina, a Chilean priest and naturalist.
Degus became popular diabetes research animals due to their inability to process sugar.
Today, degus are kept as pets in the United States, UK, and Europe. Degus are very unique and have their own quirks, so always do your research before adding these little guys to your family.

Extinction Superstar

Dodo bird

Raphus cucullatus

by Angela Boyle

On Mauritius, a small, isolated island, the dodo evolved without native predators.
Mauritius
Madagascar
N
This meant they could nest on the ground without fear.
And their food source consisted largely of ripe fruit that had fallen to the ground.
dodo tree fruit is peach-like
In this veritable paradise there was no pressure to fly, a costly endevour. With no pressure to leave the ground, that trait atrophied, leading to a sternum too weak to support flight.
collar bone
hyoid
wing
ribs
sternum

The modern image of a dodo. is often a fat, greedy bird.
Doop der doo.
But this reputation likely comes from captive birds shipped to Europe. They were unhealthy; lacking exercise and overfed.
wobble wobble
In fact, Mauritius has a wet and a dry season, which affects food availability.
So by the end of the wet season when food was still plentiful, the birds would gorge themselves.
Fr-uit!
NOM NOM NOM
The extra weight helped them survive the dry season when food was scarce.
sigh

Dutch Vice-Admiral Wybrand Van Warwijck was the first person to seemingly name the dodo bird in 1598.
We found a bird I shall call "walgvogel." So great is its natural stupidity, it can be approached and knocked down without making an attempt to escape.
Quote from Bolton's Mauritius Almanac and Official Directory, 1852
ha ha ha ha ha
booga booga
eep!
Though the origins of the team "dodo" are unclear, there is plenty of speculation. In Portugese "doidos" means "fool" or "crazy," which matches with Warwijck's (mistaken) description.
But the Portugese have no record of the dodo from their arrival on Mauritius. The Dutch maintain the first record. And in Dutch, "dodoaars" refers to the tail feathers (aars=butt).

The first time a European set foot on Mauritius was around 1505 as a stopover for a trade route.
But because it was just a quick pit stop for food and rations, it wasn't until 1598 with Warwijck that the dodo was first discovered
What the what?
Being the largest bird on the island, they were quickly killed for food.
But with humans come more vermin. Rats, monkeys, and pigs were let onto and island and made a feast of the eggs so vulnerable in the ground nests.
The last known dodo died between 1660–1685, less than 100 years after discovery. Natural philosophers were just starting to explore the world, 50 years before Linnaeus would formalize the modern naming system of species.

And the dodo wasn't all. Of the 45 species of bird discovered on the island, only 21 species remain.
Mauritius Grey White-Eye
Mauritius Kestrel
Echo Parakeet
Pink Pigeon
The destruction wasn't limited to just the direct killing of birds. There is a tree on the island (*Sideroxylon glandiflorum*) of which only 13 specimens remain
All of these trees are about 300 years old, which means no new trees have germinated since the 1600s. When the last dodo died.
Turns out, the dodo ate the fruit, and their digestive system activated the seed.
crop
cecum
proventriculus
gizzard
cloaca
Hence the name "dodo tree."
Fortunately, scientists discovered the modern domestic turkey has a similar-enough digestive tract to activate the seeds. So the tree can make a comeback.

Most of the remaining dodo skeletons are cobbled together from the bones of several dodos. Only 3 places in the world have remains from known living birds.
The Oxford University Museum of Natural History: last known stuffed dodo.
The University of Copenhagen Zoological Museum: skull from possibly the oldest remains of a live dodo in Europe.
The National Museum in Prague.
Since dodos lived in or near swamps, the swamp of Mare aux Songes has yeilded over 300 dodo fossils.
The only complete skeleton of a dodo is in the Natural History Museum at Mauritius Institute.
Across the world, 26 museums have at least part of a dodo:
·American Museum of Natural History, NY
·Cambridge University Museum of Zoology
·Senckenberg Museum in Frankfurt
See ya never!

Opossum

INVASION
OF THE
INVASIVE PLANTS
BY BONESTEEL
N E S W
THE SETTING:
RURAL AMERICA
A MODEST, THREE-BEDROOM HOUSE WITH A SIZABLE GARDEN.
DIRT-T
THIS WILL BE GROUND ZERO FOR A HORTICULTURAL NIGHTMARE THAT WILL RIPPLE OUT INTO THE COMMUNITY.

CYNOGLOSSEUM OFFICINALE

(HOUNDSTONGUE)

NATIVE TO EURASIA.

FLOWERS RANGE IN COLOR FROM RED TO BURGUNDY.

OBLONG LEAVES COVERED IN TINY WHITE HAIRS.

ONCE THE FLOWERS BLOOMED, THE OWNER RECOGNIZED THE MIX-UP BUT, KNOWING NO BETTER, LET THEM BE.
SO THEY CONTINUED TO GROW.
THE HOUNDSTONGUE TOOK OVER THE GARDEN, SPREADING LIKE WILD FIRE.
SPREADING TO ALL THE OTHER PLANT BEDS.
CHOAKING OUT ALL THE OTHER PLANTS UNTIL IT WAS THE ONLY ONE.
THE FIRST YEAR OF THE PLANT'S LIFE, IT PRODUCES NO FLOWERS. THIS OCCURS DURING THE SECOND YEAR, AFTER WHICH IT GOES TO SEED. THE SEED PODS ARE BROKEN UP INTO FOUR PART CLUSTERS. THEY CONTAIN TINY BURS THAT EASILY CLING TO LOCAL WILD LIFE AND CAN GET INTO THE EYES OF LIVESTOCK, RESULTING IN BLINDNESS.

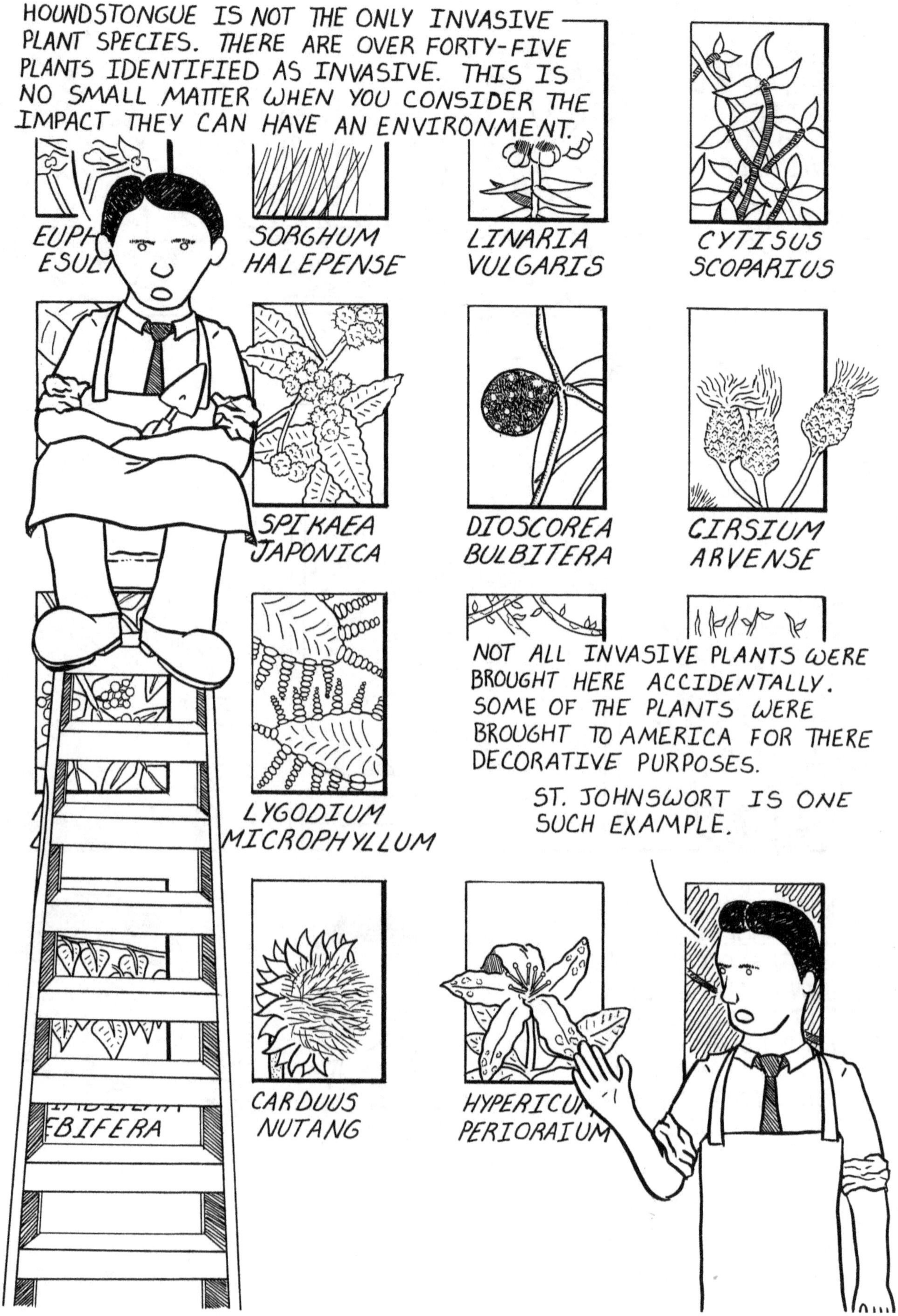
HOUNDSTONGUE IS NOT THE ONLY INVASIVE PLANT SPECIES. THERE ARE OVER FORTY-FIVE PLANTS IDENTIFIED AS INVASIVE. THIS IS NO SMALL MATTER WHEN YOU CONSIDER THE IMPACT THEY CAN HAVE AN ENVIRONMENT.
SORGHUM HALEPENSE
LINARIA VULGARIS
CYTISUS SCOPARIUS
SPIKAEA JAPONICA
DIOSCOREA BULBITERA
CIRSIUM ARVENSE
LYGODIUM MICROPHYLLUM
NOT ALL INVASIVE PLANTS WERE BROUGHT HERE ACCIDENTALLY. SOME OF THE PLANTS WERE BROUGHT TO AMERICA FOR THERE DECORATIVE PURPOSES.
ST. JOHNSWORT IS ONE SUCH EXAMPLE.
CARDUUS NUTANG

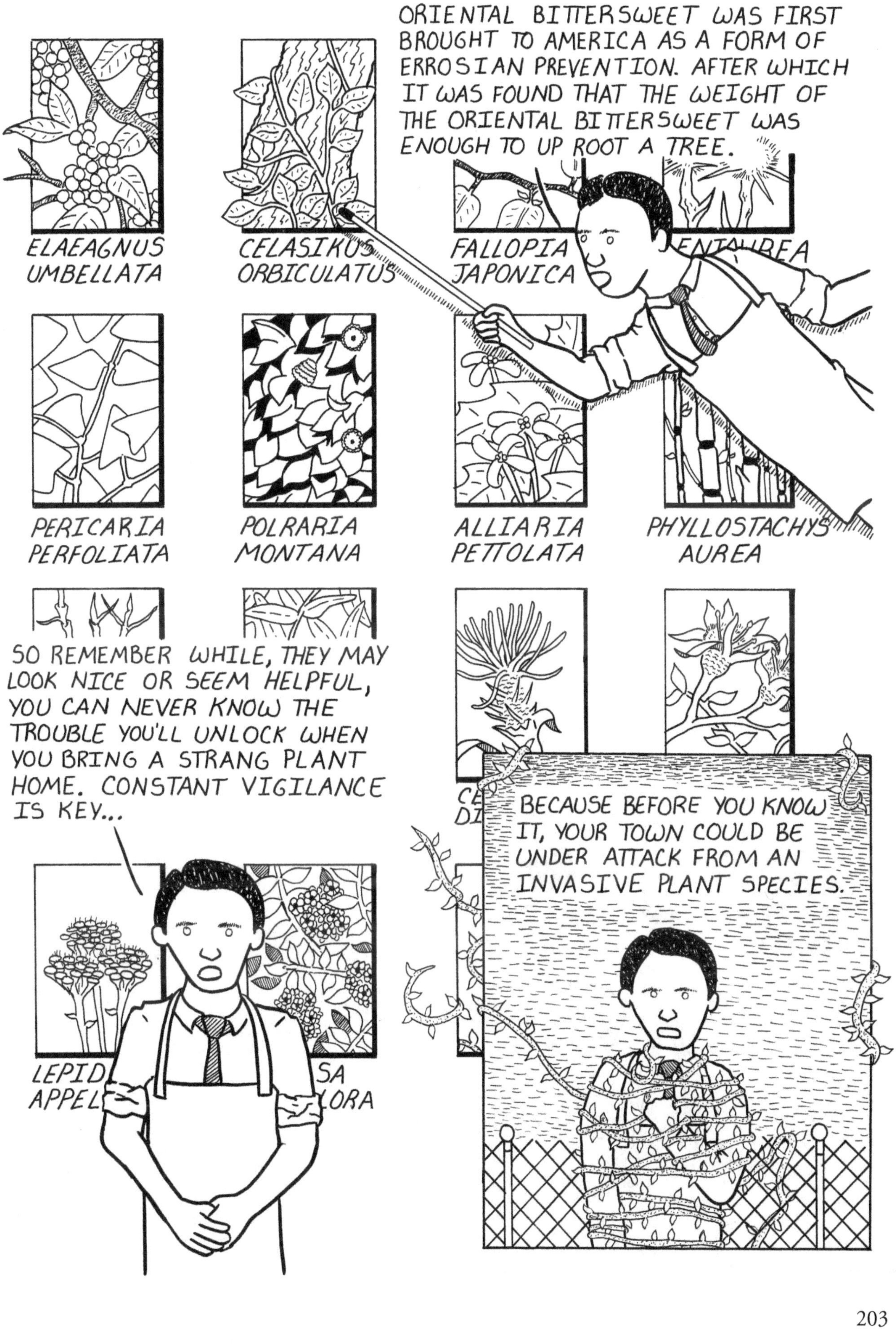
ORIENTAL BITTERSWEET WAS FIRST BROUGHT TO AMERICA AS A FORM OF ERROSIAN PREVENTION. AFTER WHICH IT WAS FOUND THAT THE WEIGHT OF THE ORIENTAL BITTERSWEET WAS ENOUGH TO UP ROOT A TREE.
ELAEAGNUS UMBELLATA
CELASIKUS ORBICULATUS
FALLOPIA JAPONICA
PERICARIA PERFOLIATA
POLRARIA MONTANA
ALLIARIA PETTOLATA
PHYLLOSTACHYS AUREA
SO REMEMBER WHILE, THEY MAY LOOK NICE OR SEEM HELPFUL, YOU CAN NEVER KNOW THE TROUBLE YOU'LL UNLOCK WHEN YOU BRING A STRANG PLANT HOME. CONSTANT VIGILANCE IS KEY...
BECAUSE BEFORE YOU KNOW IT, YOUR TOWN COULD BE UNDER ATTACK FROM AN INVASIVE PLANT SPECIES.

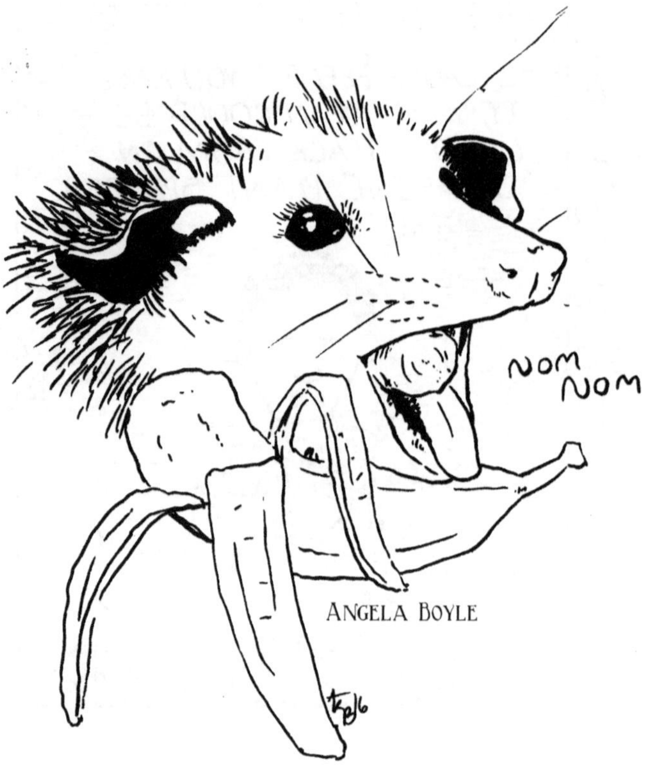

ANGELA BOYLE

Why Great Blue Herons Eat Alone

By: Kami Koyamatsu

GREAT BLUE HERONS ARE THE LARGEST HERON IN NORTH AMERICA.

THEY HUNT PATIENTLY, SITTING STILL, WAITING FOR THEIR PREY TO COME TO THEM.

THEY STRIKE WITH LIGHTNING SPEED, PLUCKING THEIR MEAL OUT OF SHALLOW WATER OR FROM AN OPEN FIELD.

GREAT BLUE HERONS TEND TO EAT FISH BUT ALSO EAT FROGS, SNAKES, INSECTS, RODENTS, AND SMALL BIRDS.

GREAT BLUE HERONS ARE ACTIVE DURING THE DAY AND NIGHT.

SOME GREAT BLUE HERONS MIGRATE, BUT THE ONES IN THE PACIFIC NORTHWEST AND IN SOUTHERN FLORIDA RESIDE YEAR ROUND.

GREAT BLUE HERON PARENTS WORK AS A TEAM TO INCUBATE AND RAISE THEIR CHICKS.

THEY TEND TO HAVE 1-2 BROODS PER YEAR WITH 3-5 EGGS IN EACH.

GREAT BLUE HERONS ROOST AND BREED IN COLONIES CALLED 'HERONRIES.'

BUT GREAT BLUE HERONS ARE SOLITARY HUNTERS AND ALWAYS EAT ALONE.

Splat
By: Kami Koyamatsu

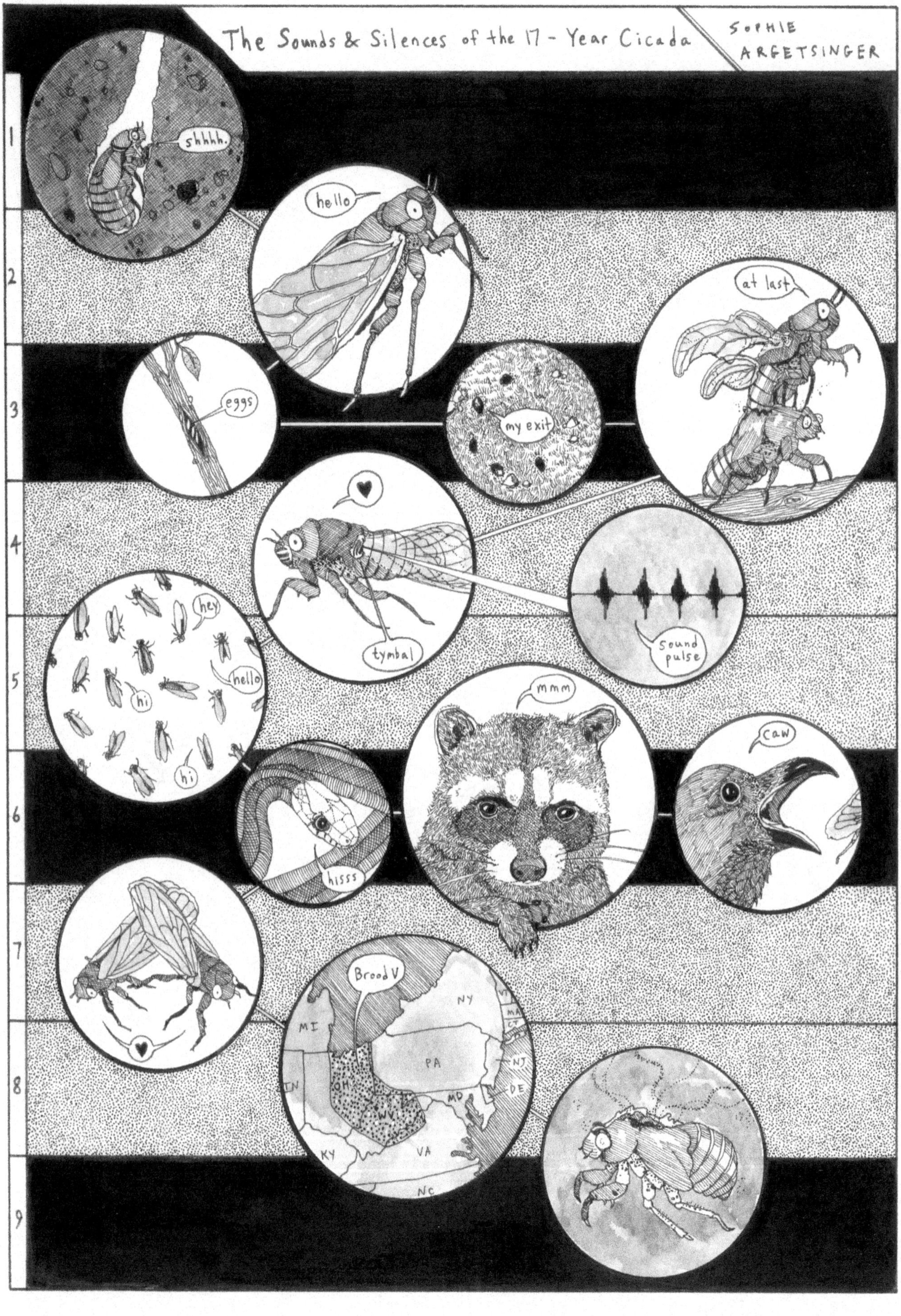
The Sounds & Silences of the 17 - Year Cicada
Sophie Argetsinger
1
shhhh.
2
hello
3
eggs
my exit
at last
4
♥
tymbal
sound pulse
5
hey
hello
hi
hi
mmm
6
caw
hisss
7
8
Brood V
MI
NY
PA
NJ
DE
MD
IN
OH
WV
KY
VA
NC
9

1 SILENCE: The 17-year periodical cicada (genus: Magicicada; species: M. septendecim, M. septendecula, M. cassini) lives most of its 17 years in silence, hidden underground.

2 SOUND: The Magicicada life cycle is completed with a timed, synchronized mass emergence from the earth. The majority of the periodical cicadas will emerge in the same year, but cicadas in different regions are not synchronized. Cicadas that emerge in a year together are known as a brood. There are 12 broods of 17-year cicadas, all located in eastern North America.

3 SILENCE: The nymphs of Magicicada are hatched from eggs the size of a grain of rice, laid in the twigs of trees. The nymphs fall to the ground, then dig themselves into the earth. They remain in their burrows, sucking the fluid from the roots of trees for 17 years, growing and shedding their skins. A few weeks before emerging, the nymphs construct exit tunnels to the surface. On a spring night, when the soil has warmed, the cicadas emerge. They find a place for themselves on nearby vegetation, complete their final molt, and rest while their exoskeletons harden.

4 SOUND: The noise begins. Male cicadas have structures called tymbals, a part of their exoskeletons. The males vibrate the membranes of their tymbals, which resonate; the sound amplifies in the large empty chambers of their abdomens. Cicadas are some of the loudest insects on earth, and males can synchronize their calls to create powerful choruses to attract females. Magicicada choruses have been documented to reach 100 decibels — as loud as a jackhammer or a jet flying 1000 feet overhead.

5 SOUND: During an emergence, tens to hundreds of thousands of cicadas per acre are common; as many as 1.5 million cicadas per acre have been observed.

6 SILENCE: Predators eat their fill of cicadas during an emergence. Periodical cicadas are slower and easier to capture than other cicadas; the species find their survival in their numbers. Predators include rodents and other mammals, birds, snakes, fish, and the cicada-killer wasp (Sphecius speciosus).

7 SOUND: Magicicada live only briefly above the ground — about four to six weeks. During this time the cicadas mate, and the females lay their eggs. Then, the adults die.

8 SOUND: In the spring of 2016, Brood V — last seen in 1999 — emerged in parts of Ohio, Pennsylvania, Virginia, West Virginia, and Maryland, and in Long Island, New York.

9 SILENCE: Now all that is left of the emergence are the dry husks of the cicadas' final molt, ghostly memories of the sounds. The hatched nymphs have dug themselves into the earth, where they grow and wait.

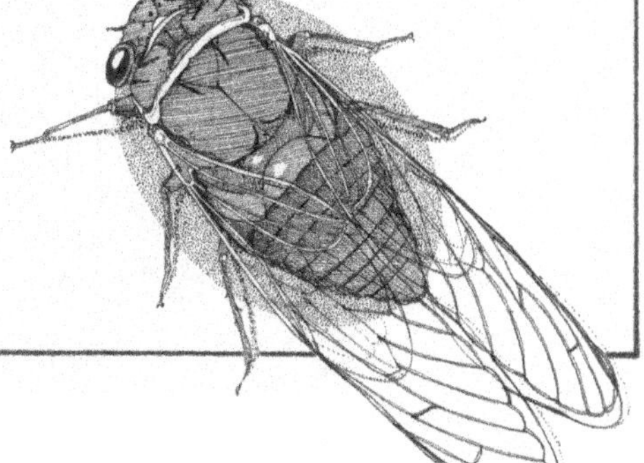

CAPTAIN 'POSSUM

BONESTEEL

THE EARS HAVE IT
How corn came to dominate America's diet
PAYDEN McROBERTS
OH HI THERE! I WAS JUST ABOUT TO ENJOY THIS #3 COMBO...
McDougal's
24
...A CORN-BURGER ON CORN SMOTHERED IN CORN SAUCE, SERVED WITH A SIDE OF CORN AND A LARGE CORN DRINK.
'BUT PAYDEN,' YOU MAY BE SAYING, 'THAT IS A CORNUCOPIA OF FOOD DIVERSITY THERE, NOT A PILE OF CORN.'
WELL, MY FRIENDS, YOU'D BE SURPRISED WHERE CORN SHOWS UP IN FOOD THESE DAYS.
LET ME SHOW YOU WHAT I MEAN.

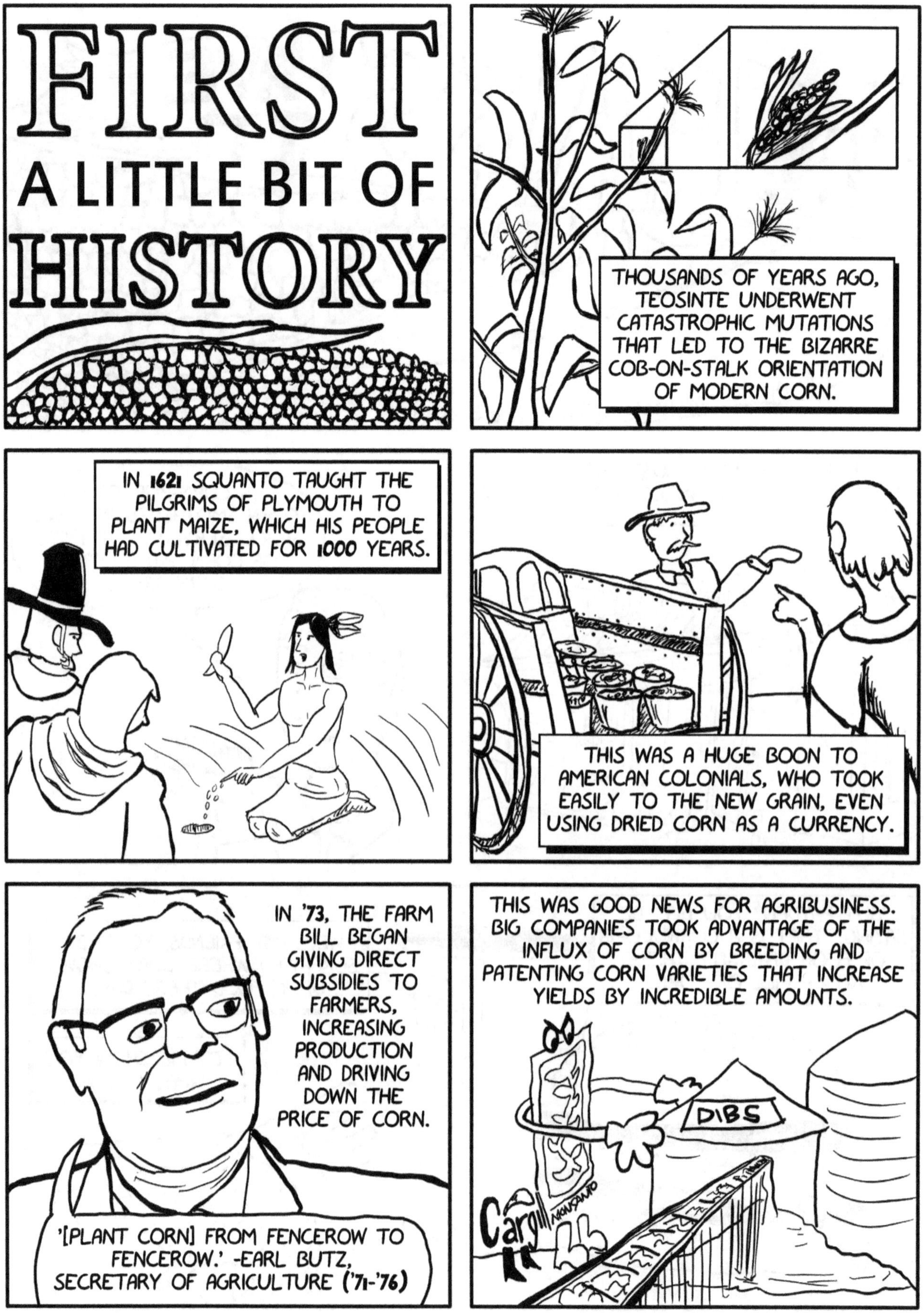
FIRST
A LITTLE BIT OF
HISTORY
THOUSANDS OF YEARS AGO, TEOSINTE UNDERWENT CATASTROPHIC MUTATIONS THAT LED TO THE BIZARRE COB-ON-STALK ORIENTATION OF MODERN CORN.
IN 1621 SQUANTO TAUGHT THE PILGRIMS OF PLYMOUTH TO PLANT MAIZE, WHICH HIS PEOPLE HAD CULTIVATED FOR 1000 YEARS.
THIS WAS A HUGE BOON TO AMERICAN COLONIALS, WHO TOOK EASILY TO THE NEW GRAIN, EVEN USING DRIED CORN AS A CURRENCY.
IN '73, THE FARM BILL BEGAN GIVING DIRECT SUBSIDIES TO FARMERS, INCREASING PRODUCTION AND DRIVING DOWN THE PRICE OF CORN.
'[PLANT CORN] FROM FENCEROW TO FENCEROW.' -EARL BUTZ, SECRETARY OF AGRICULTURE ('71-'76)
THIS WAS GOOD NEWS FOR AGRIBUSINESS. BIG COMPANIES TOOK ADVANTAGE OF THE INFLUX OF CORN BY BREEDING AND PATENTING CORN VARIETIES THAT INCREASE YIELDS BY INCREDIBLE AMOUNTS.
DIBS
Cargill
MONSANTO

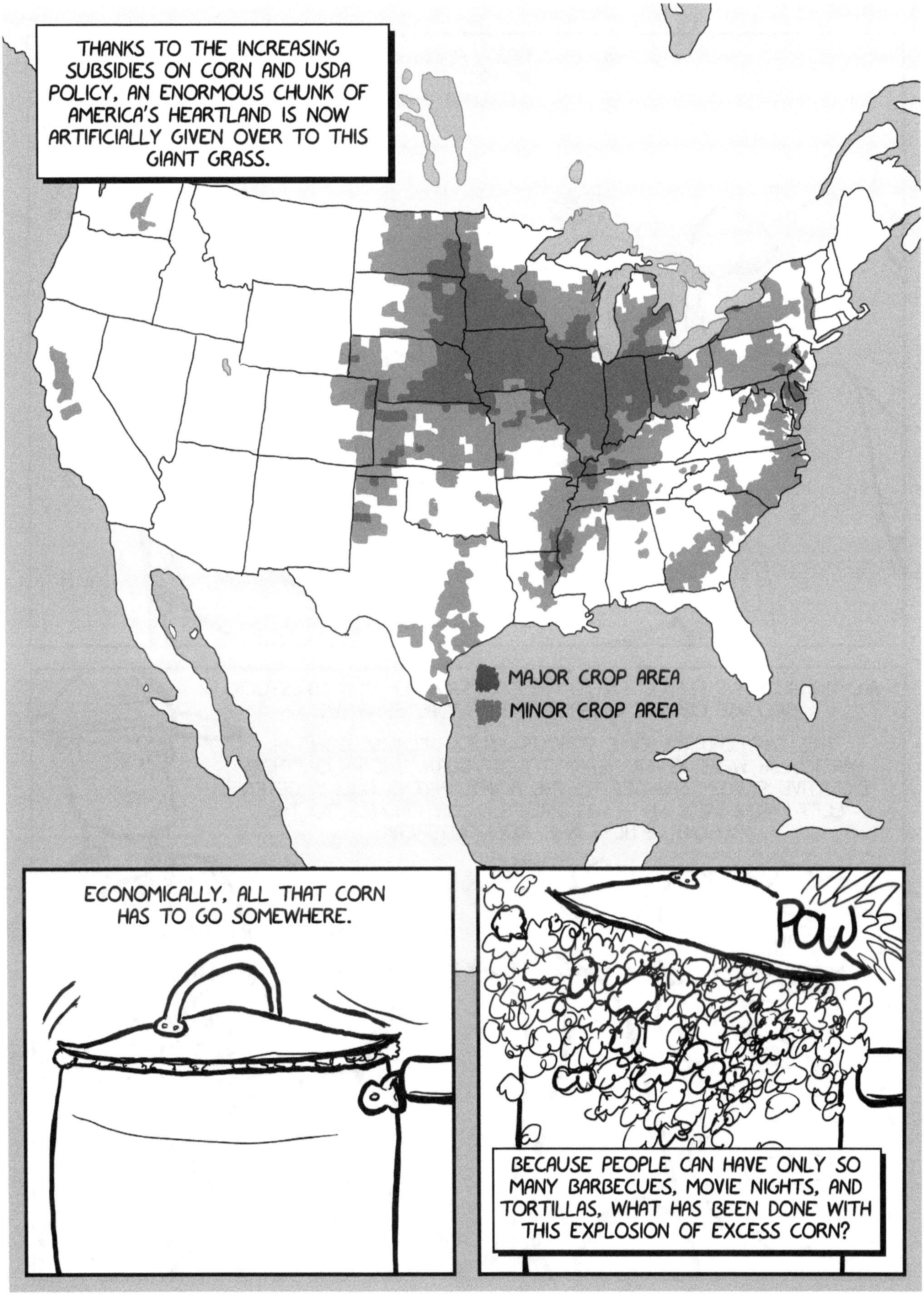
THANKS TO THE INCREASING SUBSIDIES ON CORN AND USDA POLICY, AN ENORMOUS CHUNK OF AMERICA'S HEARTLAND IS NOW ARTIFICIALLY GIVEN OVER TO THIS GIANT GRASS.
MAJOR CROP AREA
MINOR CROP AREA
ECONOMICALLY, ALL THAT CORN HAS TO GO SOMEWHERE.
POW
BECAUSE PEOPLE CAN HAVE ONLY SO MANY BARBECUES, MOVIE NIGHTS, AND TORTILLAS, WHAT HAS BEEN DONE WITH THIS EXPLOSION OF EXCESS CORN?

FIRST, FOOD SCIENCE HAS DISCOVERED A MULTITUDE OF USES FOR PROCESSED CORN. BASICALLY YOU CAN EQUATE ANYTHING ON A FOOD INGREDIENTS LABEL THAT YOU DON'T RECOGNIZE WITH CORN:
MODIFIED OR UNMODIFIED STARCH
CRYSTALLINE FRUCTOSE
GLUCOSE SYRUP
HIGH FRUCTOSE CORN SYRUP
POLYOLS
CARAMEL COLOR
XANTHAN GUM
MALTODEXTRIN
ASCORBIC ACID
LECITHIN
DEXTROSE
LACTIC ACID
LYSINE
MALTOSE
MSG

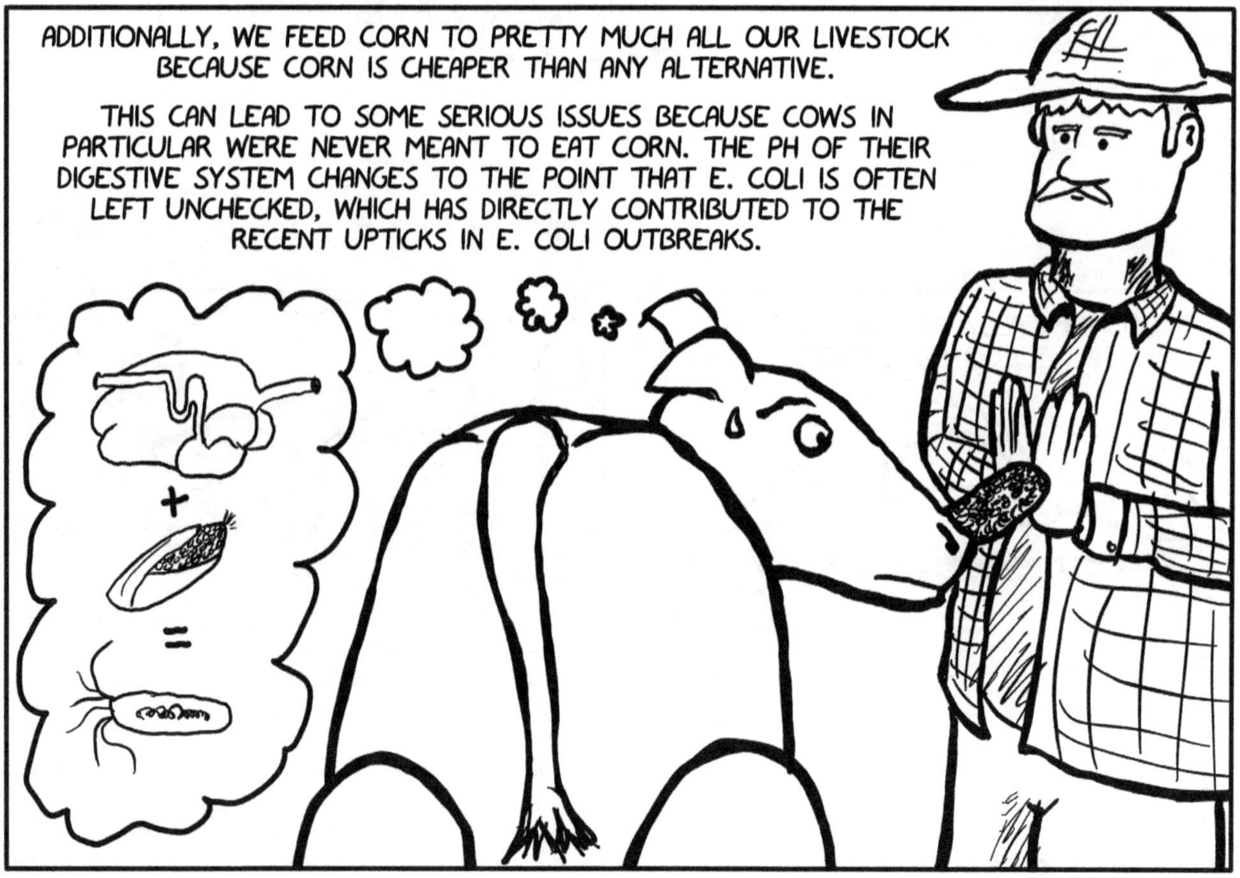
ADDITIONALLY, WE FEED CORN TO PRETTY MUCH ALL OUR LIVESTOCK BECAUSE CORN IS CHEAPER THAN ANY ALTERNATIVE.
THIS CAN LEAD TO SOME SERIOUS ISSUES BECAUSE COWS IN PARTICULAR WERE NEVER MEANT TO EAT CORN. THE PH OF THEIR DIGESTIVE SYSTEM CHANGES TO THE POINT THAT E. COLI IS OFTEN LEFT UNCHECKED, WHICH HAS DIRECTLY CONTRIBUTED TO THE RECENT UPTICKS IN E. COLI OUTBREAKS.

GIVEN THAT CHICKENS ARE VIRTUALLY ALL CORN-FED AS WELL, CHICKEN NUGGETS ARE ESSENTIALLY CORN-BATTERED CORN.
SHARK NUGGETS*
*Not made with real shark.
Dr. Paprika
PUSH
PUSH
SOFT DRINKS, THANKS TO THE LOW PRICE OF HIGH-FRUCTOSE CORN SYRUP AS A SWEETENER, CAN BE VIEWED AS CORN JUICE.
THE AISLES OF THE MARKET ARE FULL OF PROCESSED FOOD THAT ARE HARD TO PLACE ON ANY TAXONOMIC CHART.
Coffee Reame
WRENCH dressing
Ghostess
Twinkles
sponge cakes
Now with less Sponge!
YOU GUESSED IT: THEY'RE CHOCK FULL OF CORN BY ANY OTHER NAME.
EVEN SALMON, A CARNIVORE, IS BEING BRED TO FEED ON CORN.
CORN SHOWS UP IN PLACES IN THE SUPERMARKET THAT DON'T EVEN SEEM RELATED TO FOOD, SUCH AS THE GLOSS ON MAGAZINE COVERS AND THE WAX THAT GIVES PRODUCE ITS SHINE.
WEEKLY
Queen Elizabeth's NEW BABY
THE RESULTS OF THE PROLIFERATION OF CHEAP CORN ALSO EXTEND FAR BEYOND THE SUPERMARKET.

SUSCEPTIBILITY TO PESTS
ALLURE OF FARMING DECREASED
NO BIODIVERSITY ALLOWED
INCREASING CHEMICAL INPUTS
SOIL QUALITY DIMINISHED
CORN HAS ITS TENDRILS IN THE UNITED STATES' ENTIRE ECOSYSTEM BECAUSE WE HAVE SHAPED VAST SWATHS OF OUR LANDSCAPE TO FIT ITS PREFERRED HABITAT.

EEFSTEAK 2.99/lb
SODA 99¢
OUR PUBLIC HEALTH IS ALSO NOW IN THE HANDS OF CORN BECAUSE MOST OF THE PRODUCTS THAT CONTRIBUTE HEAVILY TO OBESITY, DIABETES, AND HEART TROUBLE (SOFT DRINKS, MEATS, AND PROCESSED FOODS) ARE CHEAP AND PLENTIFUL.

BILL
FINALLY, AMERICA'S FARMERS ARE AT THE MERCY OF THE ECONOMICS OF CHEAP CORN, WHICH DEMANDS THAT THEY PLANT MORE AND MORE CORN TO PAY THEIR BILLS, WHICH FURTHER DRIVES THE PRICE OF CORN DOWN.

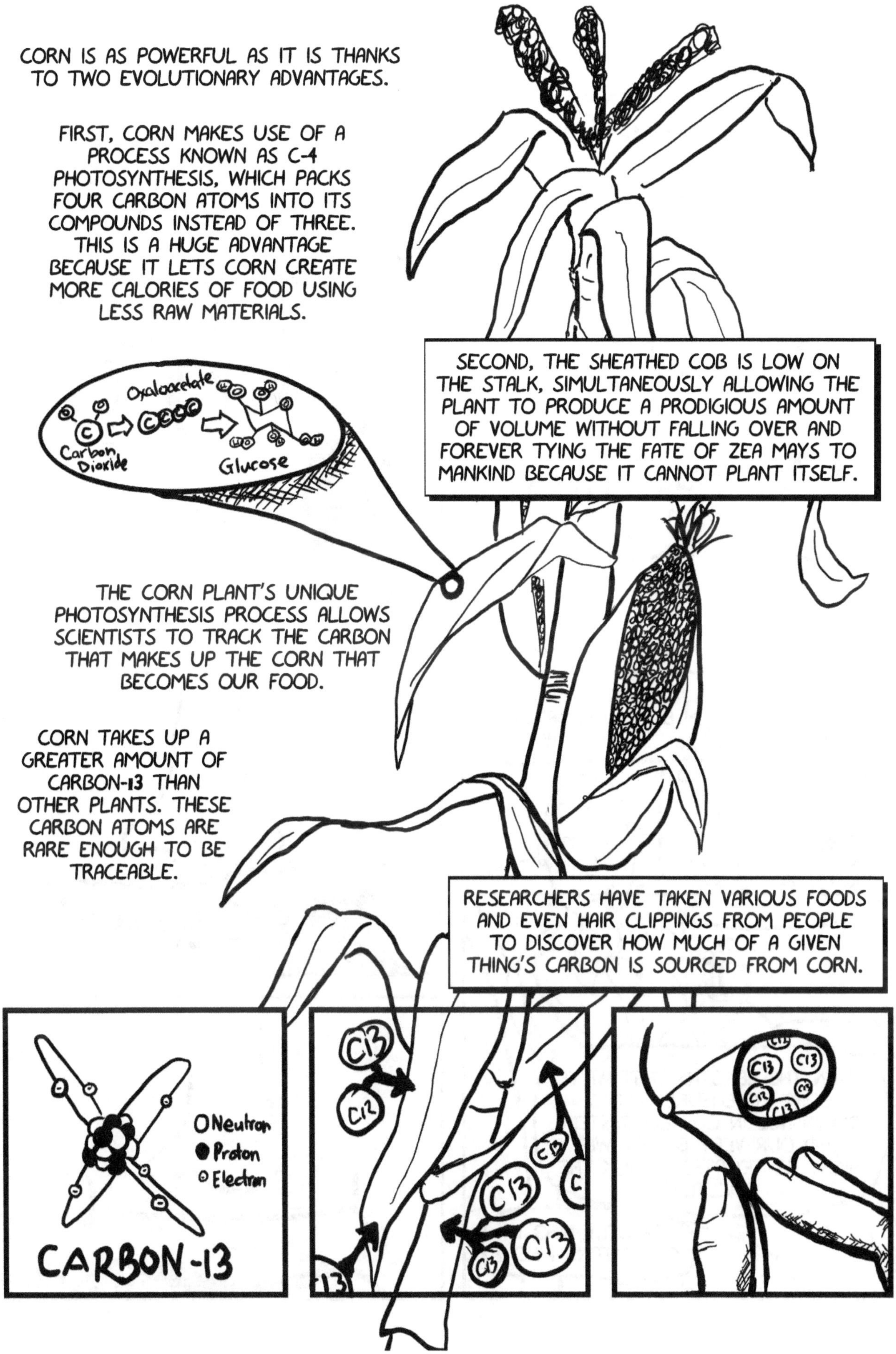
CORN IS AS POWERFUL AS IT IS THANKS TO TWO EVOLUTIONARY ADVANTAGES.
FIRST, CORN MAKES USE OF A PROCESS KNOWN AS C-4 PHOTOSYNTHESIS, WHICH PACKS FOUR CARBON ATOMS INTO ITS COMPOUNDS INSTEAD OF THREE. THIS IS A HUGE ADVANTAGE BECAUSE IT LETS CORN CREATE MORE CALORIES OF FOOD USING LESS RAW MATERIALS.
SECOND, THE SHEATHED COB IS LOW ON THE STALK, SIMULTANEOUSLY ALLOWING THE PLANT TO PRODUCE A PRODIGIOUS AMOUNT OF VOLUME WITHOUT FALLING OVER AND FOREVER TYING THE FATE OF ZEA MAYS TO MANKIND BECAUSE IT CANNOT PLANT ITSELF.
Oxaloacetate
Carbon Dioxide
Glucose
THE CORN PLANT'S UNIQUE PHOTOSYNTHESIS PROCESS ALLOWS SCIENTISTS TO TRACK THE CARBON THAT MAKES UP THE CORN THAT BECOMES OUR FOOD.
CORN TAKES UP A GREATER AMOUNT OF CARBON-13 THAN OTHER PLANTS. THESE CARBON ATOMS ARE RARE ENOUGH TO BE TRACEABLE.
RESEARCHERS HAVE TAKEN VARIOUS FOODS AND EVEN HAIR CLIPPINGS FROM PEOPLE TO DISCOVER HOW MUCH OF A GIVEN THING'S CARBON IS SOURCED FROM CORN.
Neutron
Proton
Electron
CARBON-13
C13
C12
C13
C12
C13
C13
C12

SODA
100%
HAMBURGER
52%
FRENCH FRIES
23%
TYPICAL AMERICAN
69%
WE'VE GIVEN CORN EVERY OPPORTUNITY TO THRIVE.
ECONOMICAL AND ECOLOGICAL FACTORS HAVE CONVERGED TO MAKE CORN THE LARGEST CONTRIBUTOR TO THE DIET EATEN BY MOST AMERICANS.
GIVEN THE LENGTHS THAT WE'VE GONE TO ENSURE CORN'S DOMINANCE IN OUR ECOSYSTEM AND OUR VERY BODIES, WE MIGHT BE TEMPTED TO ASK THE QUESTION:
EXACTLY WHO IS CULTIVATING WHO ?
FIN.

MONOTREMES
The egg-laying mammals
written by Abe Olson drawn by Angela Boyle
EUTHERIA — "true beasts"
Placental mammals, most common today
DOGS
CATS
BEARS
DOLPHINS
And us.
The first Eutheria appears in the fossil record around 160 million years ago.
*animals not to scale
METATHERIA — "changed beasts"
Marsupials
KANGAROO
WOMBAT
OPOSSUM
Metatheria first appear in the fossil record around 125 million years ago.
PROTOTHERIA — "first beasts"
Egg-layers, the first mammals to diverge from reptiles 220 million years ago.
My fave!
Today, there are only a few surviving species and all are members of the order Monotremata: The platypus and four species of echidna are all that remain. They live only in Australia and New Guinea.

Monotremes are distinct from other mammals in a number of ways.

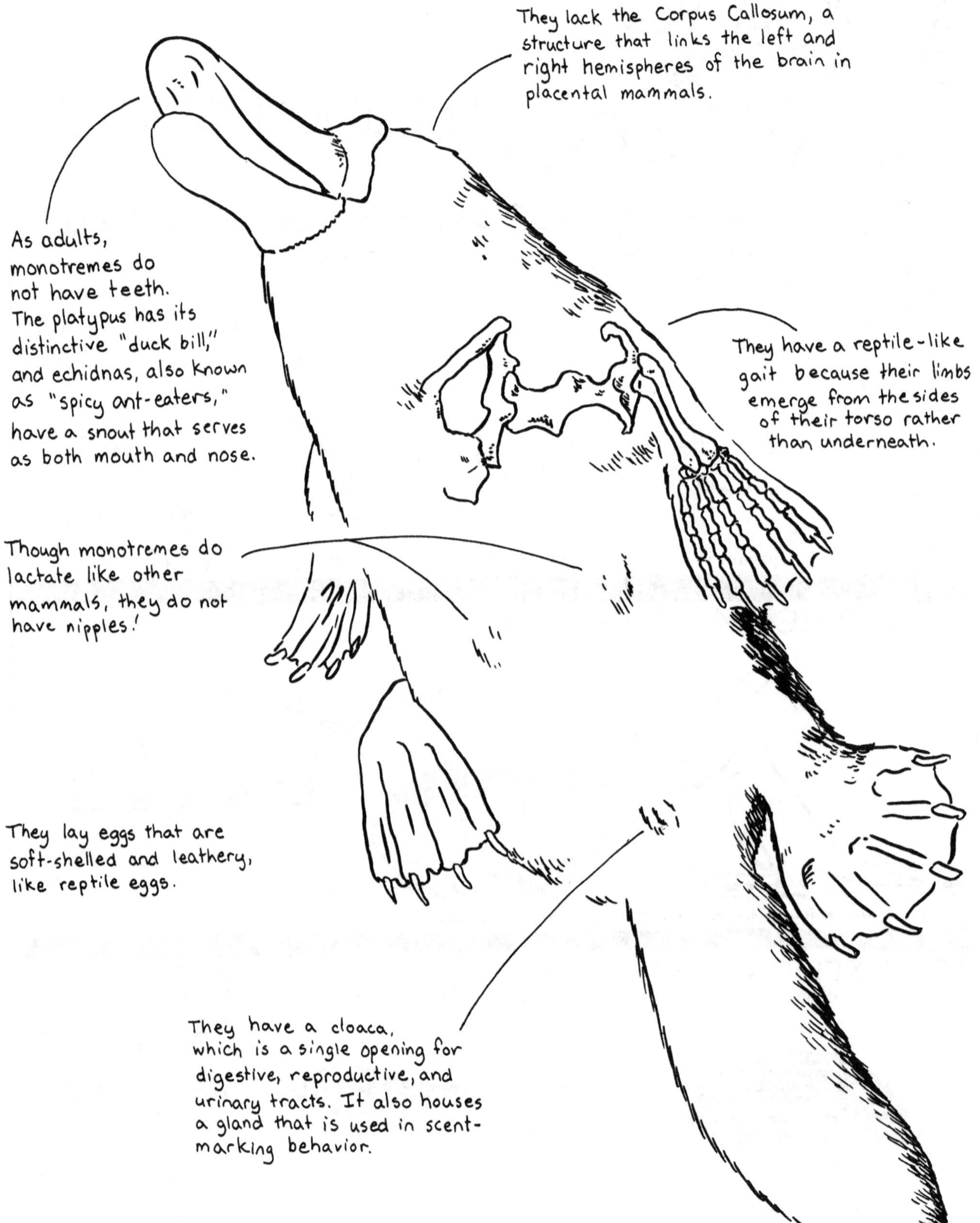

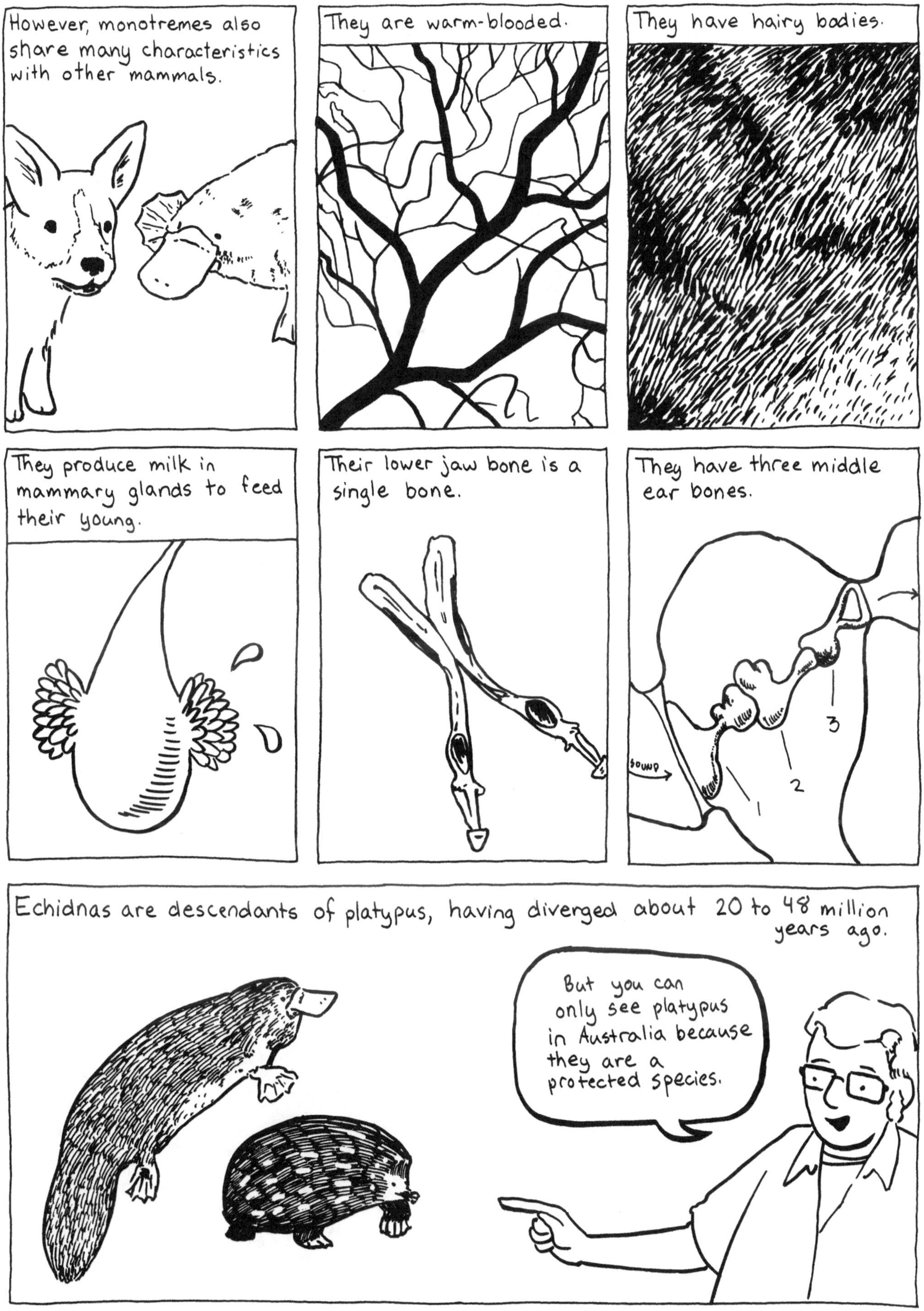
However, monotremes also share many characteristics with other mammals.
They are warm-blooded.
They have hairy bodies.
They produce milk in mammary glands to feed their young.
Their lower jaw bone is a single bone.
They have three middle ear bones.
SOUND
1
2
3
Echidnas are descendants of platypus, having diverged about 20 to 48 million years ago.
But you can only see platypus in Australia because they are a protected species.

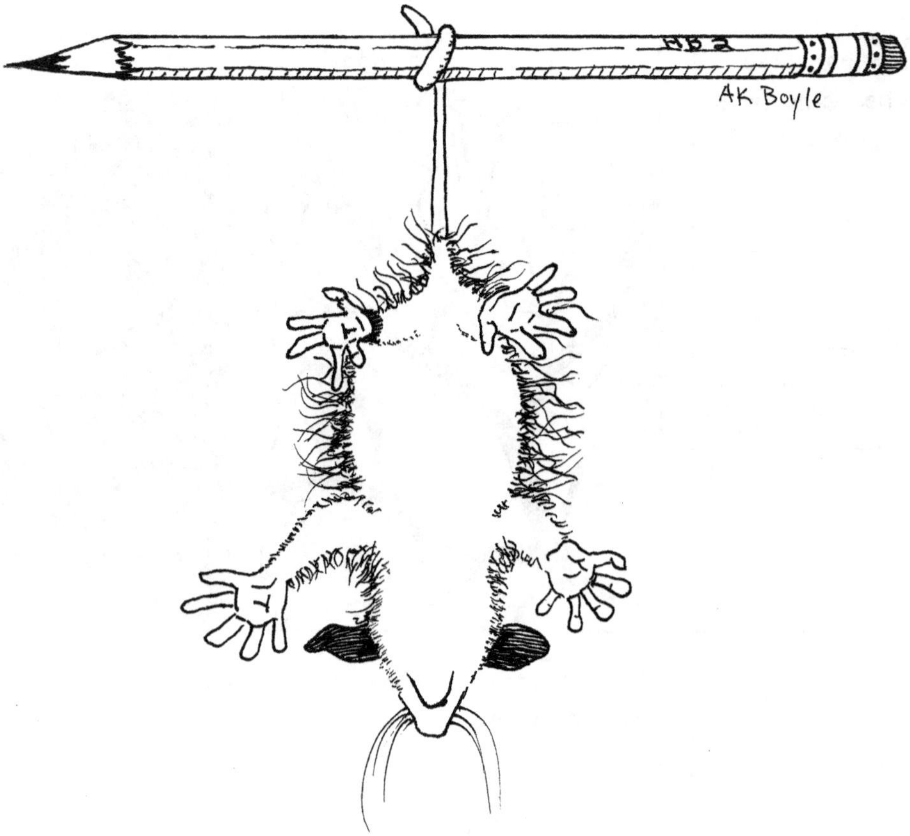
HB2
AK Boyle

COULEE CROWNS

by Laura Marie Madden

The plains cottonwood (*Populus deltoides*) is the most iconic tree of coulee ecosystems.
Plains Cottonwood
(Although it is not the only poplar to make river valleys its home.)
Black Cottonwood
Narrowleaf Cottonwood
Balsam Poplar
All poplars are dioecious— either male or female.
Male
Female
The fluffy, cotton-like seeds released from the female catkins give the cottonwoods their name.
Cottonwood cotton is a very common sight along pathways in the early summer.
Nothing is quite like cottony snowfalls in the middle of June.
...Not that these giant trees and their whimsical seeds are always appreciated.
AH!

Cottonwood trees grow fast, and they grow large.
And although they can survive drought and are a staple in semi-arid environments, they are dependent on water.
The flooding and meandering of rivers is neccesary for cottonwood trees, which need the moisture and sunlight silt and sandbars provide to germinate.
New cottonwoods will grow along the rivers, and older cottonwoods provide the shade for other species of trees to grow.
They filter water, stabilize river banks, reduce flood peaks and river speeds, and maintain river temperature.
These riparian trees are the pioneers of a diverse ecosystem that provides for many prairie residents.
Even the ones you don't usually see.

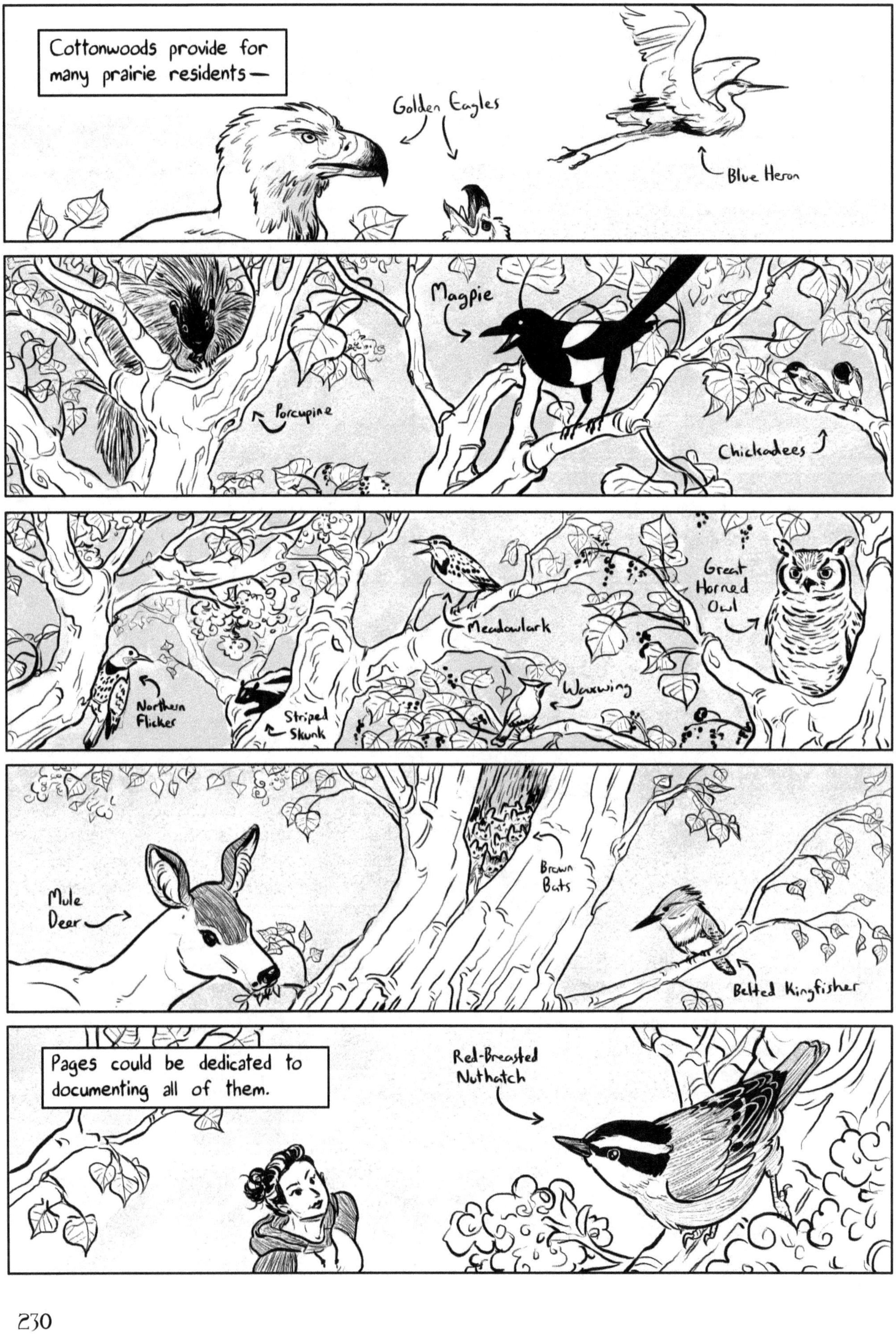
Cottonwoods provide for many prairie residents—
Golden Eagles
Blue Heron
Magpie
Porcupine
Chickadees
Great Horned Owl
Meadowlark
Northern Flicker
Striped Skunk
Waxwing
Mule Deer
Brown Bats
Belted Kingfisher
Pages could be dedicated to documenting all of them.
Red-Breasted Nuthatch

And of course, humans have plenty of uses for cottonwood forests, too!
Cottonwoods' ubiquitous presence in coulee parks goes largely unacknowledged, despite their importance to the ecosystem.
But nature never needed recognition.

...Just protection.
In Southern Alberta, cottonwood forests are iconic. And in danger.
Due to the specific conditions of germination, cottonwoods are threatened by damming projects.
Damming rivers reduces spring flooding and prevents fresh water and silt from moving as regularly as it did pre-damming.
And less water means less seedlings, which means a decline in cottonwood populations.
And less forest means less nesting birds, more erosion, less shelter from the sun, and less biodiversity.

It may not seem like much compared to the rainforests or the mountains, but prairie forests are just as valuable.
They literally hold the earth together.
With environmental stewardship, a balance between industry and nature, and renaturalization, the future of cottonwood forests can be secured.
They may not be magical to the outsider, but they are full of diverse treasures.
So next time you think about the prairies...
Think about the wonders that grow there.

ANGELA BOYLE

THE GRASSHOPPER MOUSE
BY JEREL DYE
THE AMERICAN SOUTHWEST, SUNSET
UNTIL THEY ARE OLDER,
THIS MOUSE
MUST PROVIDE,
BRAVING THE DANGERS

OF THE SCRUB DESERT
ALONE.
A BARK SCORPION,
ITS VENOM CAN KILL.

UNLIKE MOST,
THIS MOUSE
IS CARNIVOROUS
AND IMMUNE TO VENOM.

LIKE MANY PREDATORS,
IT'S TERRITORIAL
AND HOWLS
TO CLAIM THIS LAND.
JD

LIVING
LARGE
by hedj
When you hear the phrase "largest creature on Earth",
pop!
hnrrrrnnghn
hup!
BOOM!
what do you think of first?

If you thought of dinosaurs, you're halfway right! The largest land animals we know of in our planet's history were *sauropods*— herbivorous dinosaurs with long necks and tails, that walked on four legs.
!!!
The biggest sauropods known from uncontroversial evidence are titanosaurs. The largest of these is *Argentinosaurus huinculensis*.
The largest remains ever found belonged to an individual estimated to measure over 130 feet (40m) long and weigh over 80 tons (73MT)!
But titanosaurs— and most other dinosaurs*—
have been extinct for a long time.
POOF
Elephants are the largest land animal alive today. African elephants, the biggest species, can grow to weigh over seven tons (6.35MT) and reach 13 feet (4m) in height. Elephants also have the longest gestation period among mammals— almost 2 years!
plink
But elephants are far from being the largest organisms on Earth.
FLIK!
*There are dinosaurs still living on Earth today; we call them birds.

PLOP
By weight, the current record holder for biggest animal— past or present— is the blue whale.
Adult female blue whales can reach lengths of over 80 feet (24m) and weigh almost 150 tons (136MT).
Like other rorqual whales, the blue whale has deep grooves running the length of its throat, allowing it to expand massively.
Huge amounts of water and food are taken in, then strained back out through the baleen plates in the mouth, leaving the food behind.
The largest non-mammalian animal living today (by combined weight and length) is the whale shark.
The longest whale shark measured was just under 40 feet (12m) long and weighed over 23 tons (21MT); sightings of even larger individuals have been reported but not confirmed.
Like blue whales, whale sharks are filter feeders and pose no threat to swimming humans.
So, are blue whales the largest living things on our planet?
PLASH!
Not by a long shot.

The true giants of our world aren't measured in feet.
Instead, they're measured in acres.
In Utah's Fishlake National Forest is a colony of quaking aspen trees known as "Pando," which is Latin for "I Spread." Pando is a *clonal* colony; not only are its tree stems genetically identical, but they are all part of the same organism and are assumed to share a single root system.
Pando's area has been measured at 106 acres, and it's estimated to weigh about 6600 tons (5987MT)— that's over 13 million pounds (5.9mil. kg.) Pando is also one of the oldest living things on Earth— its root system is believed to be around 80,000 years old.
But even Pando can't claim to be the largest known organism by area; that title goes to a fungus.
OREGON
A specimen of *Armillaria solidipes*, formerly *A. ostoyae*, was found in 1999 to have spread over approximately 2500 acres in Malheur National Forest, OR.
Its mushrooms all grow from the same fibrous network, which spreads through the ground and makes up the bulk of the organism.
This fungal mat is estimated to be much younger than Pando, at "only" 2500 years; it's thought to weigh less than a third of what Pando does too.

To find the biggest living things on Earth, you don't need to look in the ocean or in the fossils of the past.
Instead, just look right under your feet!

Kelly Swann

Monsters Unmasked

By Patricia Maldonado

Fear can distort the way we see the world. It alters our perceptions and has the power to change a normal event into a nightmare.

The slaughter of hundreds of victims starting in 1764 caused widespread panic in the province of Gevaudan in southern France. Victims were found with their throats ripped out, and the gruesome slayings could be explained only by a fearsome beast with supernatural strength.

At the time, France and the rest of Europe were plagued by wolves who would often attack humans in packs. There were thousands of deaths in the 1700s alone.

The glimpse of a figure unknown . . .

In the late 1990's, el chupacabra terrorized Latin America, slaughtering livestock and draining them of their blood. The creature was described as having reptile-like skin and a ridge of sharp spines along its back.

Coyotes afflicted with mange often appear to have dark, leathery skin and protruding spinal ridges. A sickly coyote unable to hunt might target pent up livestock.

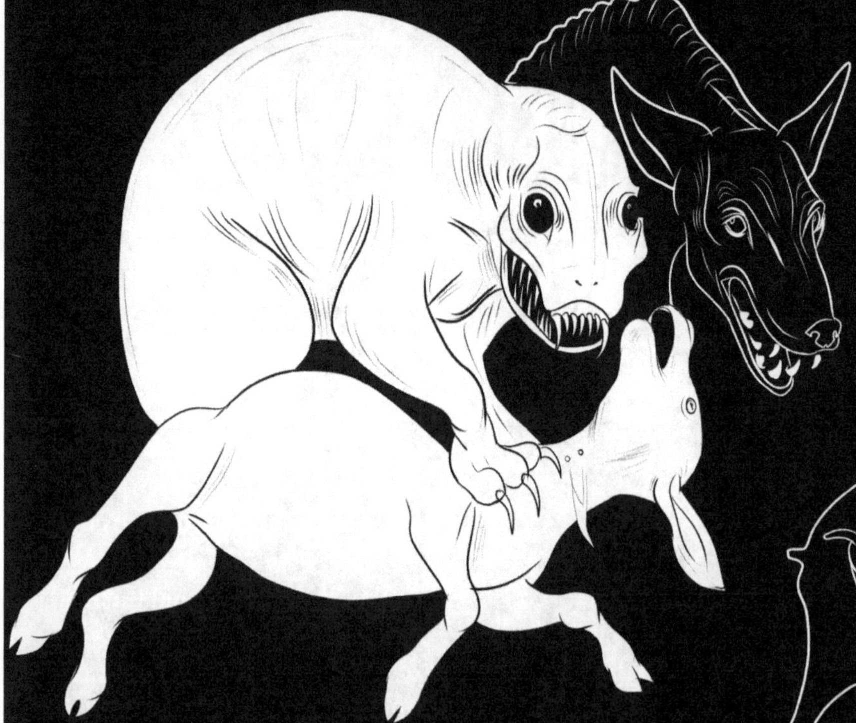

The legends of the Jersey devil date back to the 1730s. It is said to terrorize the pine barrens of southern New Jersey with its blood-curdling scream.

Although uncommon, the sandhill crane can be spotted in the marshes of New Jersey and is known for its long call that can be heard over long distances.

A howling shriek in the night...

. . . are more than enough to create a monster.

VICKY YUH
Filial Cannibalism
happens when a parent of an animal species eats some or all of its own offspring, or the offspring of another in their species.
What kinds of animals participate in this? Both males and females, from insects to birds to mammals and more. They can be influenced by different factors, even among their own species.
I'm an assassin bug, and I have practiced filial cannibalism.
So have I! I'm a house finch.
Filial cannibalism, "FC" for short, is considered a bit of an evolutionary conundrum because there are many mixed reasons why parents eat their young. The patterns observed have not all been universal but sometimes even contradictory.
I had my reasons...
The main factors contributing to FC can be categorized into: animal personality, environment, and individual plasticity.
Individual plasticity is the individual's adaptability to environmental changes or differences.

Overall, studied cases of goby fish show that FC is influenced not only by environment variation but also activity, "a behavior with consistent individual differences." Different activities may or may not be directly related to breeding.

Not so well... I haven't been able to find much food for myself.
At this rate, I might have to eat some or all of my eggs.
Well I've been on fire! My babies are developing just fine, and there's nothing more I can do for them, but...
The water feels hotter these days, and it's got me all energized!
I need to do something... anything!
Oh boy... we all know what that means. With all the activity spillage, you might end up eating your eggs, too!
It's true! My metabolism has risen from this minor increase in water temperature!
But your eggs are already in an advanced stage of development!
They're healthy, too. I would just leave them be if I were you.

In your case, if you have to eat all your eggs, do it. I'm sure you can make an even better quality brood in the next breeding cycle.
It's better than if you were to be off foraging and then have someone else come to snatch them up.
Thanks for the support, guys. It's hard to find other gobies who won't turn around and eat your clutch.
Daww... So, any news on the current about other gobies?

Well, I heard that this one goby has way too many eggs in his nest right now.
I'm pretty sure he wants to get rid of some of them.
Oof. Yeah. Overcrowding can have some problematic consequences. Just wait until they all hatch.

And since he's found another chance to mate, he might eat all of the current eggs, depending on their quality.
I wonder if the new female had already raided his nest, before approaching him?
That's a high possibility. It makes sense to get the obstacle out of the way first. But this goby has so many eggs! Think she could've manage...
WOOSH
WOOSH
WOOSH
—OK. Enough chit chat. Let's get back to our respective clutches.
See ya!
NOM
NOM
NOM

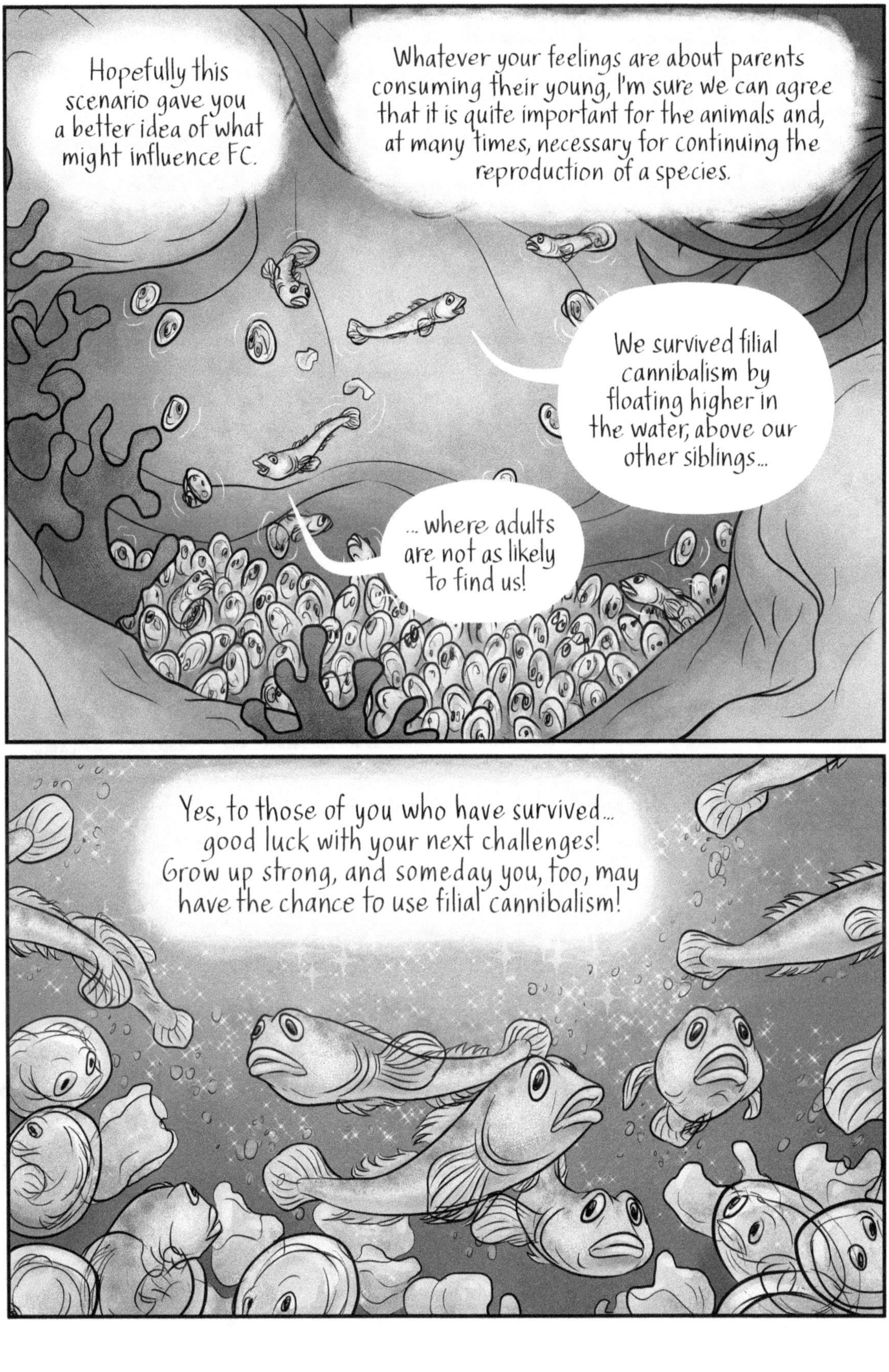
Hopefully this scenario gave you a better idea of what might influence FC.
Whatever your feelings are about parents consuming their young, I'm sure we can agree that it is quite important for the animals and, at many times, necessary for continuing the reproduction of a species.
We survived filial cannibalism by floating higher in the water, above our other siblings...
...where adults are not as likely to find us!
Yes, to those of you who have survived... good luck with your next challenges! Grow up strong, and someday you, too, may have the chance to use filial cannibalism!

MATING SUPPORT
GROUP
Dr. Catkins
AURORA MELCHOR
AND THEN, WHEN HE HAD... WELL... FINISHED...
I... BIT HIS HEAD OFF

MISS MANTIS, THANKS FOR SHARING YOUR EXPERIENCE.
IT TAKES COURAGE TO BE THE FIRST ON THE FLOOR.
CAN I ADD TO THIS?
SURE! MISS REDBACK SPIDER.
I HAVE DONE THAT TOO!
YES, SEXUAL CANNIBALISM IS QUITE WIDESPREAD AMONGST INSECTS & ARACHNIDS.
IT'S A PURRFECT EXAMPLE TO OPEN WITH.
LET ME INTRODUCE YOU TO...
Sexual Conflict

LIFE HASN'T ALWAYS BEEN THE WAY IT IS. IT CHANGES. IT CHANGES. IT EVOLVES. ACCORDING TO DARWIN:
BROADLY SPEAKING, LIVING BEINGS ARE SHAPED BY THE STRUGGLE TO SURVIVE.
BUT THEY ARE ALSO SHAPED BY THE STRUGGLE TO REPRODUCE.
THIS IS WHAT WE KNOW AS NATURAL SELECTION.
THIS WE KNOW AS SEXUAL SELECTION.
AND IT CAN EXPLAIN WHY MR. PEACOCK HAS EVOLVED TO HAVE SUCH EXUBERANT COLORS. LOOK AT HIM! I MEAN, HOW CAN THAT TAIL HELP HIM TO SURVIVE? IT'S LIKE AN OPEN INVITATION TO GET EATEN!
THERE MUST BE ANOTHER FACTOR AT WORK. AND THAT IS SEX.
PEAHENS SEEM TO PREFER MALES WITH BRIGHT FEATHERS, SO THEY WILL PREFERENTIALLY MATE WITH FANCIER DUDES. OVER TIME, FEATHERS BECOME INCREASINGLY FLAMBOYANT, EVEN IF THAT MAKES MALES AN EASY PREY.
THAT IS, LIVE FAST, DIE YOUNG!!

SO IT IS NOT ENOUGH TO SURVIVE TO PASS ON YOUR GENES TO THE NEXT GENERATION.
1
2
3
YOU MUST BE CHOSEN TO REPRODUCE.
BUT FEMALES AND MALES OFTEN HAVE CONFLICTING REPRODUCTIVE STRATEGIES, AND THIS CREATES A CONSTANT TUG-OF-WAR.
SEXUAL CONFLICT
CONFLICT OVER HOW OFTEN, WHEN, AND WITH WHOM TO MATE. CONFLICT OVER PARENTAL CARE. CONFLICT COMES IN MANY FORMS.
MRS. REDBACK HAS GIVEN US A GREAT EXAMPLE.
THE MALE WANTS TO MATE. BUT SHE DOESN'T. HE'S NOT GOOD ENOUGH FOR A PARTNER.
BUT HE'S GOOD ENOUGH FOR A SNACK.
SEE? HE JUST WASN'T "THE ONE"!
NOW I'D LIKE TO INVITE YOU TO SHARE YOUR EXPERIENCES!
WHO'D LIKE TO SPEAK?

YES, MR. COMMON SIDE-BLOTCHED LIZARD?
THERE IS A TRAITOR IN MY FAMILY!
THE TRAITOR WAS PART OF MY HAREM, ONE OF MY WIVES. UNTIL YESTERDAY.
WHEN I DISCOVERED SHE WAS A GUY!
AH, THE FASCINATING LIFE HISTORY OF SIDE-BLOTCHED LIZARDS!
BASED ON THE COLOR OF THE NECK, THERE ARE THREE TYPES OF MALES:
RED SPOT
ULTRA-DOMINANT
PHYSICALLY SUPERIOR TO THE OTHER MALES. KEEPS A HAREM OF FEMALES, WHICH HE GUARDS WITH TOOTH AND CLAW.
BLUE SPOT
DOMINANT
KEEPS ONE FEMALE AS PARTNER, EASY TO GUARD AGAINST REDS WHO'D TRY TO STEAL HER.
YELLOW SPOT
SNEAKY
PHYSICALLY WEAKEST. PRETENDS TO BE A FEMALE AND SNEAKS INTO HAREMS TO COPULATE WITH FEMALES.

HANG ON…
THAT MEANS…
UUM, YEAH, YOUR CHILDREN MIGHT NOT BE REALLY YOURS.
DR. CATKINS!
YES, MR. PENDULINE TIT?
I'M CONCERNED THAT MY PARTNER HASN'T LAID ANY EGGS.
WE MET EARLY THIS SEASON AND… NOTHING.
SHE HAS BEEN LAYING EGGS ALRIGHT. SHE JUST THROWS THEM AWAY SECRETLY, SO YOU, A LESSER BUT LOYAL MALE WILL HELP HER RAISE THE OFFSPRING OF A MORE PROMISCUOUS MALE.
PARENTHOOD CAN BE A TRICKY THING.

IN SEXUAL CONFLICT ALL KINDS OF SUBTERFUGE HAPPEN.
WE'VE HEARD ABOUT DISGUISE, DESERTION...
THOUGH, IN MOST CASES, IT IS THE FEMALE WHO GETS DITCHED.
I WISH I WAS DESERTED! HE'S BEEN THERE FOR AGES...!
IN ORDER TO MAKE SURE OF THEIR PATERNITY, WATERSTRIDERS WILL COPULATE FOR A LONG TIME.
OFTEN MONTHS.
JEEZ, GEORGE! I WISH YOU LASTED THAT LONG!
WHAT ABOUT ME, DR.? HE'S NOT EVEN FROM MY SPECIES!
WELL, SOME MALE FROGS WILL HUMP ANYTHING THAT DOESN'T HUM, WHICH FEMALES DO WHEN THEY DON'T WANT TO MATE.
THEY'LL HUMP ANYTHING...

DR., I AM... UHM... CONFUSED
I WAS HANGING OUT WITH MY BEST MATE, JUST CHILLING, WHEN HE TOOK OUT HIS PENIS AND STABBED ME WITH IT!
AND NOW I'M A GIRL.
AH. YOU'RE A SEA SLUG!
THEY ARE HERMAPHRODITIC. MEANS EACH INDIVIDUAL HAS BOTH SEXES.
WHEN THEY COPULATE, SEA SLUGS FIGHT TO DETERMINE SEX ROLES. THEIR PENISES ARE SHARP, AND WHOEVER GETS STABBED FIRST BECOMES THE FEMALE.
IT'S CALLED PENIS-FENCING.
ANY MORE QUESTIONS?
RUB
RUB
Doctor
YES?

Finally, finished!
Hey Catkins, check out my new Tinder date!
THE END (For Now)

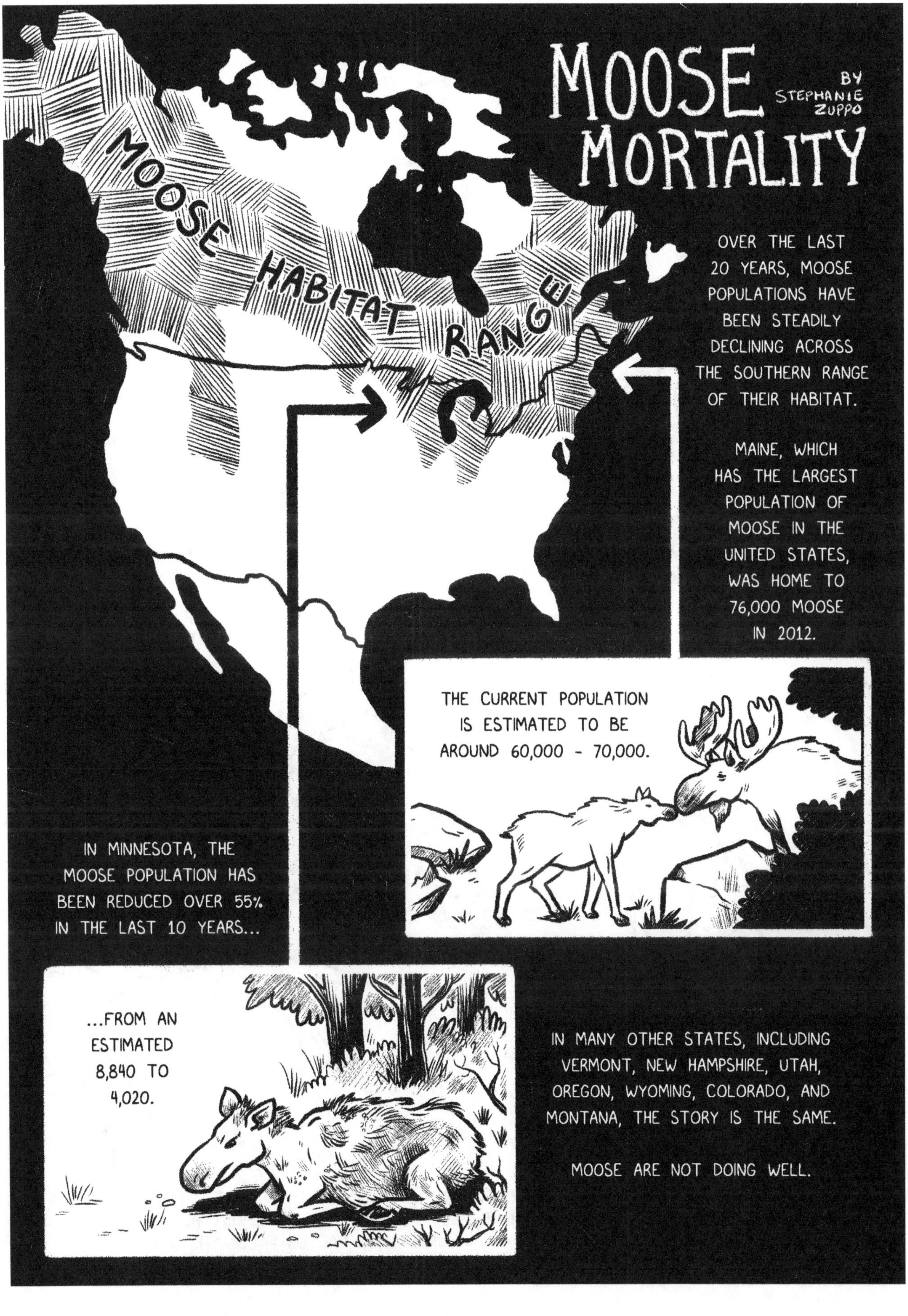
MOOSE MORTALITY
BY STEPHANIE ZUPPO
MOOSE HABITAT RANGE
OVER THE LAST 20 YEARS, MOOSE POPULATIONS HAVE BEEN STEADILY DECLINING ACROSS THE SOUTHERN RANGE OF THEIR HABITAT.
MAINE, WHICH HAS THE LARGEST POPULATION OF MOOSE IN THE UNITED STATES, WAS HOME TO 76,000 MOOSE IN 2012.
THE CURRENT POPULATION IS ESTIMATED TO BE AROUND 60,000 - 70,000.
IN MINNESOTA, THE MOOSE POPULATION HAS BEEN REDUCED OVER 55% IN THE LAST 10 YEARS...
...FROM AN ESTIMATED 8,840 TO 4,020.
IN MANY OTHER STATES, INCLUDING VERMONT, NEW HAMPSHIRE, UTAH, OREGON, WYOMING, COLORADO, AND MONTANA, THE STORY IS THE SAME.
MOOSE ARE NOT DOING WELL.

RESEARCHERS ARE ONLY JUST BEGINNING TO UNDERSTAND WHAT MAY CAUSE THE INCREASING NUMBER OF MOOSE DEATHS EACH YEAR.

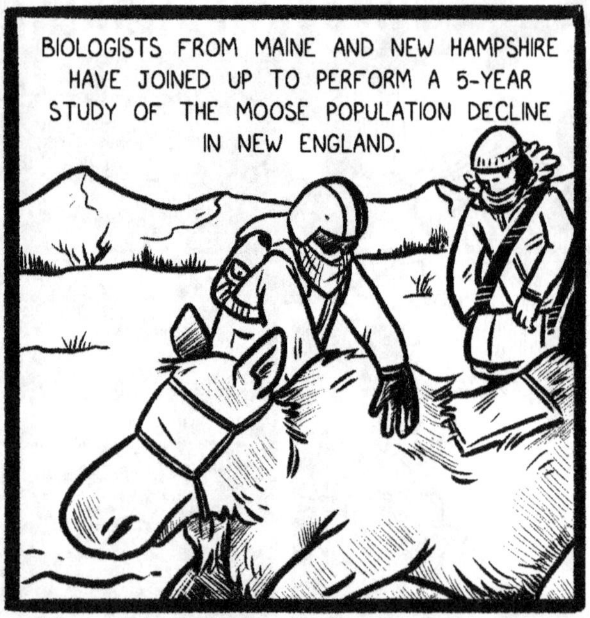

WHEN THEY RECEIVE A MORTALITY NOTIFICATION, THEY FIND THE MOOSE AND PERFORM A BIOPSY TO BETTER UNDERSTAND THE CAUSE OF DEATH.

BETWEEN THE TWO STUDIES, VARIOUS CAUSES OF DEATH HAVE BEEN DETERMINED, BUT ALMOST ALL INVOLVED ONE UNDERLYING FACTOR - CLIMATE CHANGE.

CLIMATE CHANGE HAS BROUGHT ABOUT A VARIETY OF HEALTH ISSUES TO MOOSE POPULATIONS.

NOT ONLY ARE THERE LESS MOOSE CALVES BEING BORN, BUT A GREATER NUMBER OF CALVES ARE NOT SURVIVING THE FIRST TWO YEARS OF THEIR LIVES.

THE MORTALITY RATE OF TAGGED CALVES WAS 74% IN THE MAINE AND NEW HAMPSHIRE STUDY LAST YEAR. AN OVERLOAD OF WINTER TICKS WAS CITED AS THE MAIN CULPRIT.

WARMER WINTERS AND LONGER SUMMERS HAVE CAUSED A SUDDEN FLOURISH OF PARASITES - SOME OF WHICH MOOSE HAVE NEVER ENCOUNTERED BEFORE.

WINTER TICKS

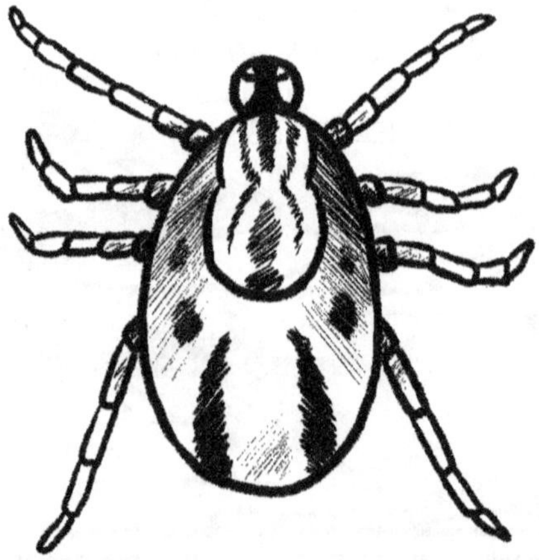

THESE TICKS ARE USUALLY KEPT IN CHECK BY DYING IN HEAVY SNOWFALL. HOWEVER, WITH LESS SNOW, THEY CAN SURVIVE THE COLD, LEADING TO AN EXPLOSION OF THE WINTER TICK POPULATION IN RECENT YEARS.

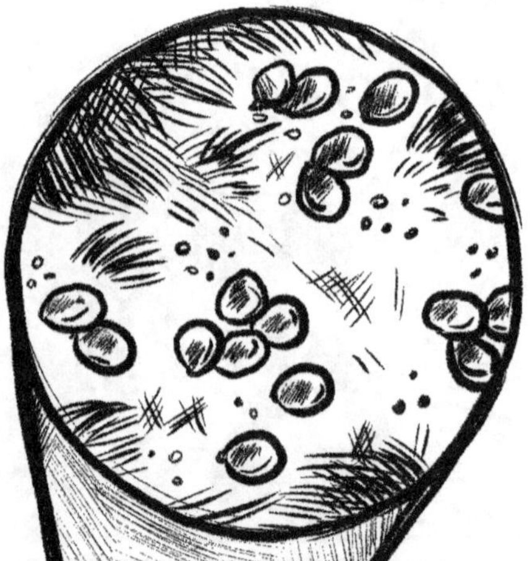

INDIVIDUAL MOOSE IN NEW HAMPSHIRE HAVE BEEN FOUND TO BE INFESTED WITH OVER 150,000 WINTER TICKS. SUCH A HIGH NUMBER OF TICKS RESULTS IN ENOUGH BLOOD LOSS TO LEAVE THEM ANEMIC AND WEAKENED TO OTHER PARASITES AND DISEASES.

MOOSE TRY DESPERATELY TO REMOVE THE TICKS BY RUBBING OFF WHOLE PATCHES OF HAIR. THESE "GHOST MOOSE" ARE A RESULT OF RUBBING OFF SO MUCH FUR THAT THEIR WHITE UNDERCOAT SHOWS THROUGH.

BRAIN WORM

MOOSE ARE THE ACCIDENTAL HOSTS OF THIS PARASITE, WHICH PRIMARILY INFECTS WHITE-TAILED DEER.

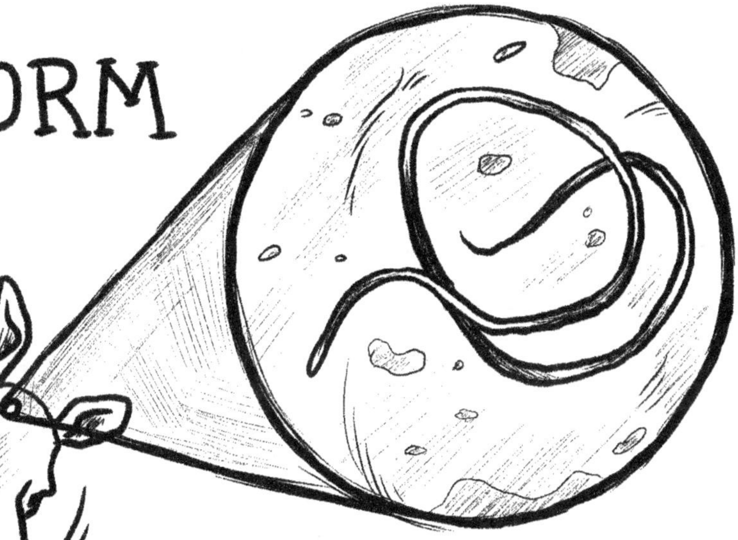

AS TEMPERATURES GRADUALLY INCREASE, DEER HABITATS ARE MOVING UPWARD INTO MOOSE TERRITORY, INTRODUCING THIS DEADLY PARASITE.

WHILE DEER AND BRAINWORM CAN LIVE TOGETHER WITH ALMOST NO ILL EFFECTS, MOOSE SHOW A NUMBER OF SYMPTOMS INCLUDING:

EXTREME WEAKNESS
CIRCLING
BLINDNESS
LETHARGY
LOSS OF FEAR
RAPID EYE MOVEMENT

A MOOSE INFECTED WITH BRAINWORM MAY SEEM ZOMBIE-LIKE OR UNPREDICTABLE -

WALKING AROUND IN CIRCLES OR STANDING IN ONE PLACE UNTIL THEY DIE OR BECOME EASY PREY TO WOLVES.

OTHER, LESS DESTRUCTIVE BUT STILL DANGEROUS, PARASITES INCLUDE -
LUNGWORM
COMMON IN MAINE; CAN MULTIPLY RAPIDLY IN THE LUNG AND LEAD TO DECREASED LUNG CAPACITY.
LIVER FLUKE
CAN CAUSE CYSTS AND INFECTION IN MOOSE LIVER AND MIGRATE ELSEWHERE IN THE MOOSE'S BODY.
CAROTID ARTERY WORM
TRANSMITTED BY HORSEFLYS; CAN CAUSE BLINDNESS, RESTRICTION OF CIRCULATION, MALFORMED ANTLERS, AND CROPPED EARS AND NOSES.
ALTHOUGH MOOSE POPULATIONS ARE SUFFERING IN WARMER AREAS, IT IS NOT LIKELY THAT THEY WILL BE DISAPPEARING ENTIRELY ANY TIME SOON.
MOOSE CONTINUE TO THRIVE IN THE COOLER CLIMATES OF NORTH AMERICA, WITH APPROXIMATELY ONE MILLION LIVING IN CANADA AND TWO HUNDRED THOUSAND IN ALASKA.
END.

The Immortal Jellyfish

by Kelly Chen

Instead of dying from old age or sickness...
through a process called transdifferentiation, the jellyfish reverts back to its immature polyp state.
From the Polyp state, the immortal jellyfish begins it's life again!
unlike most animals, in favourable conditions the cycle of life goes on indefinitely for the immortal jellyfish!

FLIGHTLESS RAILS AND ISLAND CONSERVATION

by G Pike

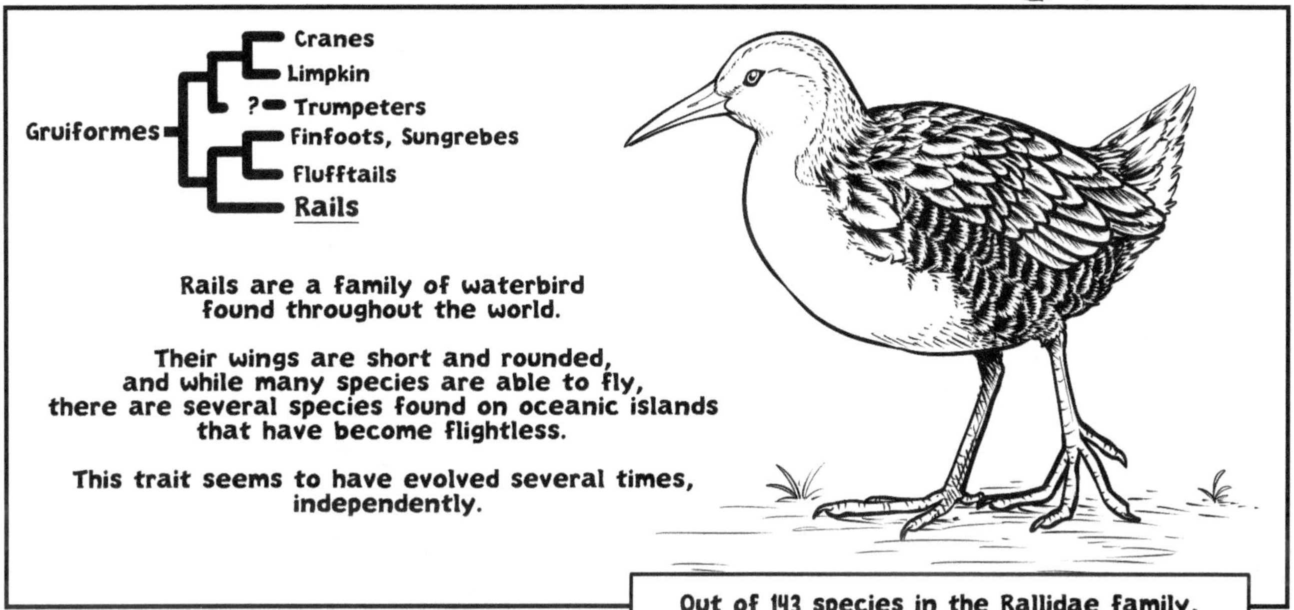

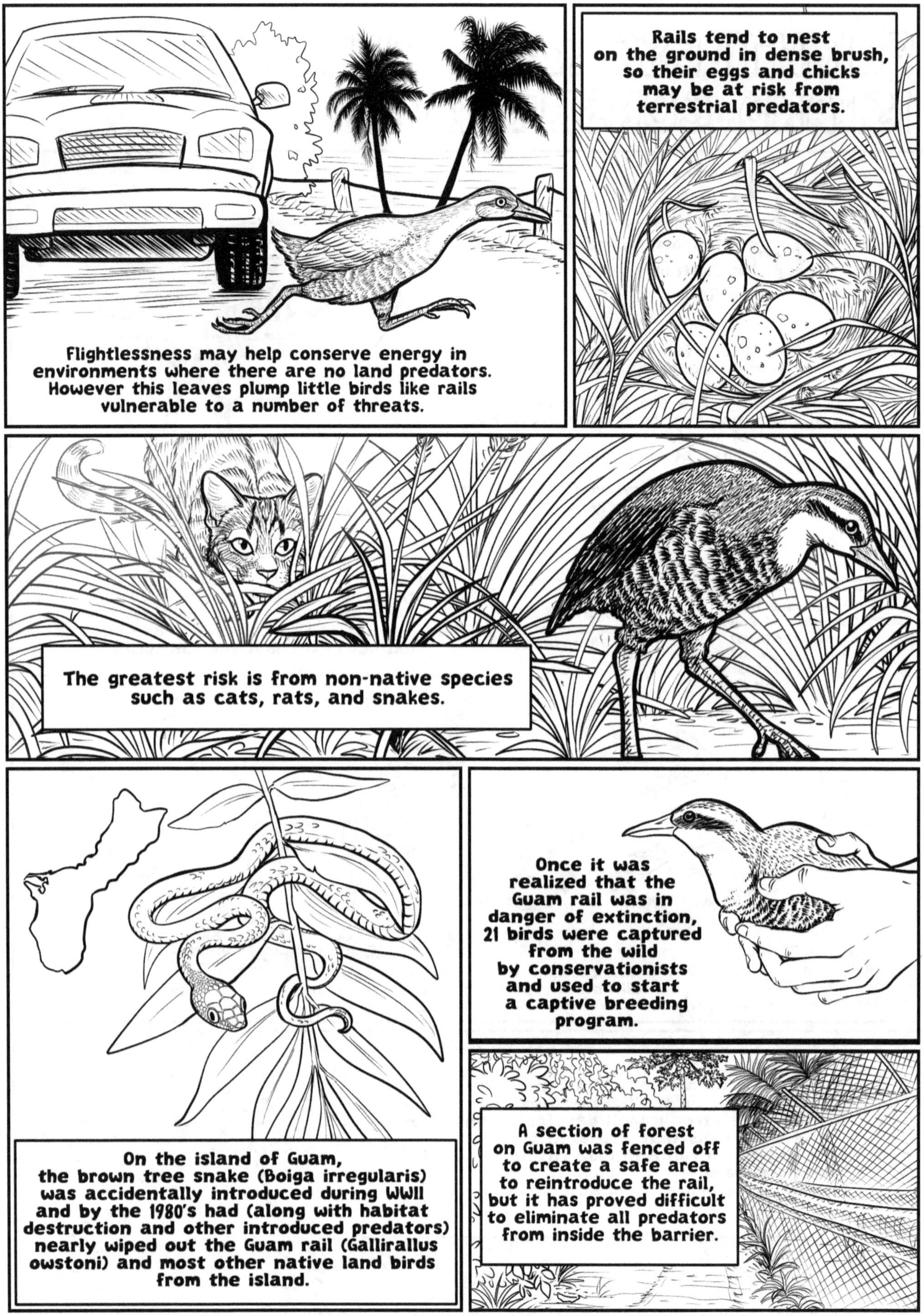
Flightlessness may help conserve energy in environments where there are no land predators. However this leaves plump little birds like rails vulnerable to a number of threats.
Rails tend to nest on the ground in dense brush, so their eggs and chicks may be at risk from terrestrial predators.
The greatest risk is from non-native species such as cats, rats, and snakes.
On the island of Guam, the brown tree snake (Boiga irregularis) was accidentally introduced during WWII and by the 1980's had (along with habitat destruction and other introduced predators) nearly wiped out the Guam rail (Gallirallus owstoni) and most other native land birds from the island.
Once it was realized that the Guam rail was in danger of extinction, 21 birds were captured from the wild by conservationists and used to start a captive breeding program.
A section of forest on Guam was fenced off to create a safe area to reintroduce the rail, but it has proved difficult to eliminate all predators from inside the barrier.

So far, the most successful reintroduction efforts have involved releasing Guam rails to smaller, nearby islands where no snakes are present.
Elsewhere in the pacific, the Wake Island rail (G. wakensis), which was numerous in the early 20th century and able to coexist with the Pacific rat (Rattus exulans), was reportedly driven to extinction by habitat destruction and hungry Japanese soldiers during WWII.
In New Zealand, the gigantic takahē (Porphyrio hochstetteri) was believed extinct until 1948, when it was rediscovered in a remote mountain region.
Humans have been trying to save the takahē for decades - but the threat of invasive predators such as stoats and competition for food from introduced deer are still very real problems.
New Zealand's other flightless rail, the weka (G. australis), is also threatened by invasive predators and habitat loss. Weka are omnivorous and sometimes accidentally fall victim to poisons or traps designed to control pests.
The future for flightless rails and thousands of other unique species of island wildlife around the world is uncertain. Conservation is urgent, but funding and awareness for these projects is often lacking. The only hope is for enough people to make an effort to protect the environment before it's too late.

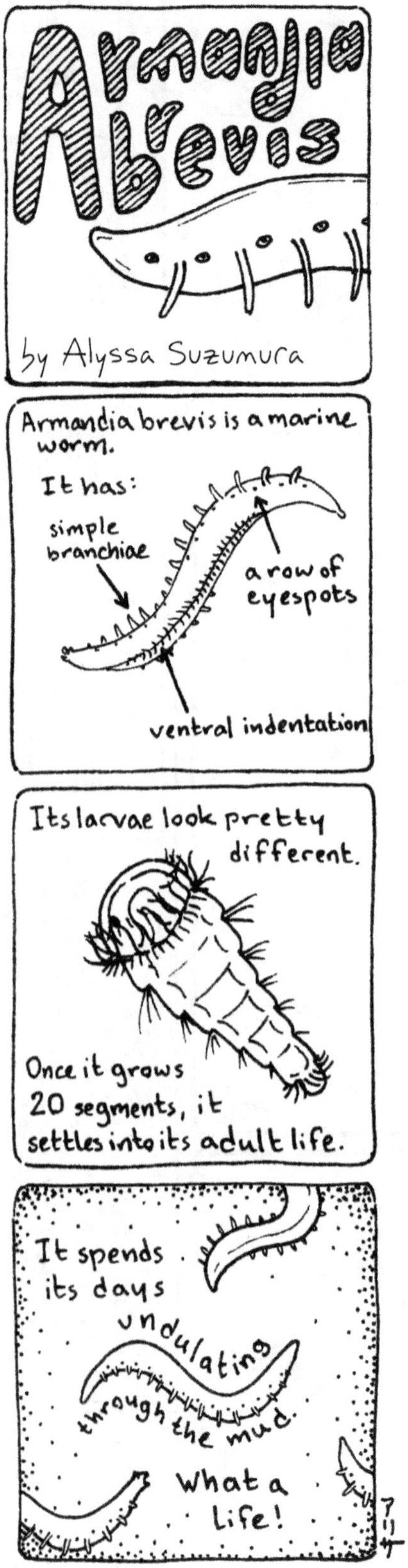
Armandia Brevis
by Alyssa Suzumura
Armandia brevis is a marine worm.
It has:
simple branchiae
a row of eyespots
ventral indentation
Its larvae look pretty different.
Once it grows 20 segments, it settles into its adult life.
It spends its days
undulating
through the mud.
What a life!

How I met your father
by Iris Yan
Mama! Mama! Can you please tell us a story? Pleease??
Hmm...ok.
DAD #1
DAD #2
DAD #3
DAD #4
Where should I start from?
Start from our dad first!
Your dad was very nice. He brought me a gift when we first met.
It was wrapped up, and it looked like a baby! So I went to get it quickly and got hooked!
Ah! Is this for me?
Yes, yes.

Were there really baby spiders inside the gift?
No. But by the time I found out, it was already too late.
What about MY dad? Did he bring a gift, too?
Yes, he did. But it was an actual fly, not a fake baby. Your father also pretended to be dead.
What?!
He just lay there with his gift in his hands so I thought he was dead.
What did you do?
Nothing. We mated.
Mom, what about MY dad?
Oh. HIM. May he rest in peace.
What happened to him?
I ate him up.
MOM!
THE END

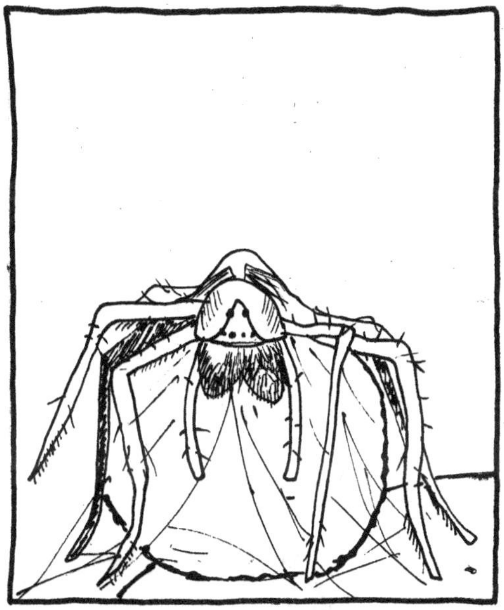

The occurrence of nuptial gifts is rare in spiders, well known in only a single species, the nursery web spider (*Pisaura mirabilis*). The males offer females prey wrapped in silk during courtship, though some males can mate without an offering. Some also exploit the maternal instinct by producing a gift that closely resembles the female egg sac.

Males of the spider *Pisaura mirabilis* also engage in spectacular mating displays involving thanatosis (death feigning). Thanatosis in a sexual context is exceptional and suggested to function as an antipredation strategy toward potentially cannibalistic females. In an experiment, death-feigning males were more successful in obtaining copulations and gained longer copulations.

Sexual cannibalism is when one partner eats the other after sex. For example, the female spider *Argiope bruennichi* tries to grab and wrap up the male at the onset of mating so she can prey on him during sex. In the lab, only about 30% of the males survive their first mating, but by letting the female gnaw on them, the males prolong the sex act, making it more likely that they will inseminate their partner.

Meet Percy!

Percy's here to help us learn about declawing.

Percy's family adores him, but they can't read his body language very well. Although Percy may ask them "politely" to stop a behavior, they don't notice until he asks much more firmly, with his claws. Because of his humans' inability to understand his needs they think he is lashing out aggressively, and may opt to declaw him to prevent it.

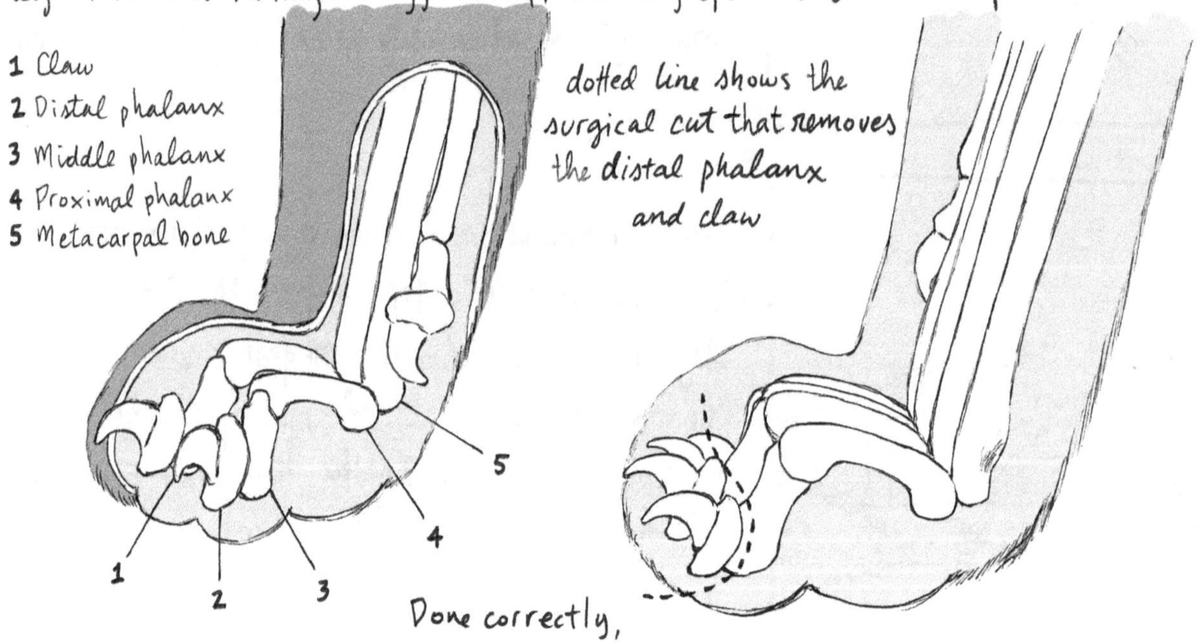

Done correctly,

The standard declawing procedure (or onychectomy) removes not just the claw but the last digital bone, preventing the claw from regrowing.

If they were to do this, Percy may be uncomfortable even after recovering from surgery. In the absence of his last digital bone, his weight would be resting on his middle phalanx. In addition, since Percy relied on his claws to communicate his needs and preferences, he may now resort to biting to protect himself against unwanted behavior.

Fortunately, Percy's humans talked to their vet, who pointed them towards a variety of non-surgery alternatives, including becoming more familiar with Percy's body language so that they can better understand and respect him.

CAT BODY LANGUAGE

Understanding your cat's body language can do so much to improve your relationship. Since cats can't just tell us what they're thinking, it's up to us to interpret their signals as best we can. The following are some examples of ways your cat may try to let you know how they're feeling.

Tall ~ hello! Tall body, slightly arched back, with upright tail is a greeting and a solicitation for attention, especially when accompanied by head-butting, rubbing their body on you, and purring or chirping.

Belly up ~ play with me! Although some cats do enjoy a belly rub, they tend to be in the minority. This presentation is more likely a solicitation to play. Remember, when cats play with each other, there are a lot of claws & teeth involved, so it's a good idea to meet this invitation with a toy, rather than your hands.

Slow blink. A slow blink or half-lidded eyes are a sign of relaxation. If they're doing it around you, they likely feel very safe & relaxed. Try returning the expression & see if they repeat it.

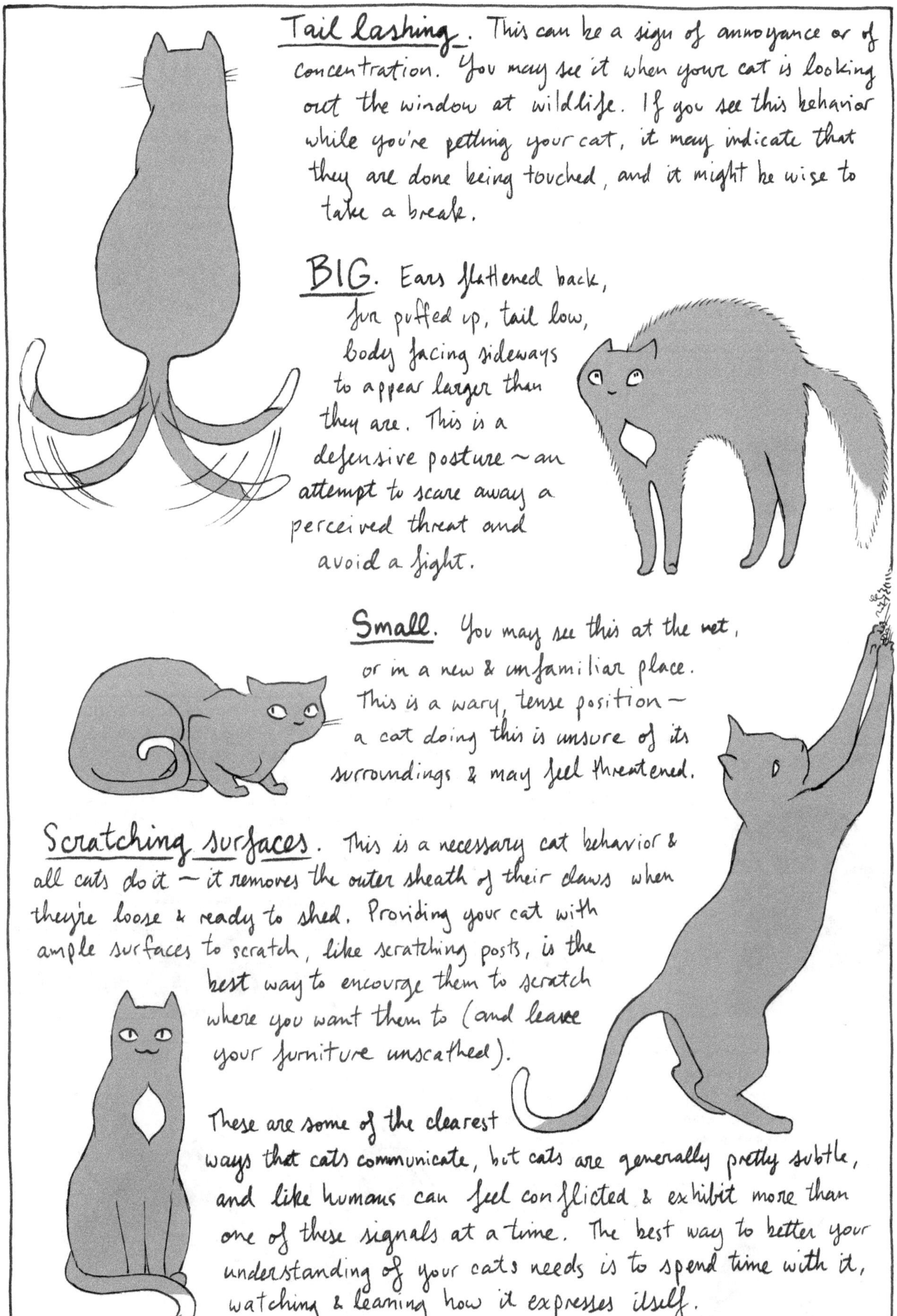
Tail lashing. This can be a sign of annoyance or of concentration. You may see it when your cat is looking out the window at wildlife. If you see this behavior while you're petting your cat, it may indicate that they are done being touched, and it might be wise to take a break.
BIG. Ears flattened back, fur puffed up, tail low, body facing sideways to appear larger than they are. This is a defensive posture ~ an attempt to scare away a perceived threat and avoid a fight.
Small. You may see this at the vet, or in a new & unfamiliar place. This is a wary, tense position ~ a cat doing this is unsure of its surroundings & may feel threatened.
Scratching surfaces. This is a necessary cat behavior & all cats do it ~ it removes the outer sheath of their claws when they're loose & ready to shed. Providing your cat with ample surfaces to scratch, like scratching posts, is the best way to encourge them to scratch where you want them to (and leave your furniture unscathed).
These are some of the clearest ways that cats communicate, but cats are generally pretty subtle, and like humans can feel conflicted & exhibit more than one of these signals at a time. The best way to better your understanding of your cats needs is to spend time with it, watching & learning how it expresses itself.
·ejl·
EJ LANDSMAN 2016

CUTE OR DEADLY: Starlings
by Angela Boyle & Abe Olson
A flock of starlings is called a murmuration, which can be up to a million individuals.
Starlings are not native to North America. In the 1890s, Eugene Schieffelin released sixty starlings in Central Park, NY. As president of the American Acclimation Society, he tried to introduce every bird species mentioned in the works of William Shakespeare. Into North America.
They have beautiful iridescent coloring.
It took just 60 years for the birds to arrive in the Pacific Northwest. There are now over 200 million of them living in North America.
2,858 miles
(4,600 km)
And they are reknowned for their ability to mimic music and the human voice.
Starlings are responsible for over 50% of airplane bird strikes in North America.
On October 4, 1960, a flock of about 20,000 starlings flew into the path of flight 375, which had just taken off from Logan Airport in Boston.
3 of 4 airplane engines were destroyed, the airplane rolled over and crashed into the water just 600 feer from the end of the runway.
62 people died. This is the worst accident in airline history caused by a bird strike.
Mozart had a pet common starling that could sing part of his Piano Concerto in G Major (KV. 453) He bought it from a shop after hearing it sing a phrase from a work he wrote six weeks previously and had not yet been performed in public. He became very attached to the bird and arranged an elaborate funeral when it died three years later.
Seems to be...
more Deadly than cute.

I'm a
fox
Wm Brian MacLean
roostertree.ca
I have kits to feed,
and so I hunt.
When old enough to wander, my kits will be more prone to predation by bobcats, hawks, and coyotes than I am as an adult.
similar to fishermen's bait-pulling technique
But in my search for food, I endure many close encounters with threats to my life.

Some threats are from the natural world...
... some less so.
My small size helps me navigate human habitats while avoiding human activity.
Our relatives, the coyotes, tend not to den near humans.
me, downwind
Due to our mutual dislike, I avoid coyotes unless my breeding den is threatened.

*Latin; so old that the Gospel of Luke alludes to it

Farmers set out poisoned bait for those larger canids; fox poisoning deaths are largely unintentional.
Some nuisance foxes are killed, sure. But we foxes are, in fact, often helpful to farmers.
vole
We hunt pestilence-vector vermin like mice and rats that also pillage stores of cereals and grain.
literally called "mouse leap"

* level of risk after endangered   † disputed

Some farmers raise us for slaughter. Fox fur was stylish during the 1920s and 30s, but this fad dropped off after WWII.
Renewed demand for fox fur came in the 1970s. Nearly 700,000 pelts were harvested in North america during the 1977-78 season alone.
By 2000, our hides had passed from vogue.

O Humanity! with your actions based on economics, nostalgia, and guilt — all notions of self-interest.
SAN MIGUEL ISLAND
LOS ANGELES
< 200 km
For example, look at California's Channel Islands.
SAN CLEMENTE ISLAND
The US Navy actively kills one subspecies of grey fox on San Clemente
endangered San Clemente loggerhead shrike
while less than 200 km north on San Miguel
golden eagle
a different subspecies dwindled to just six individuals before humans began efforts to repopulate.

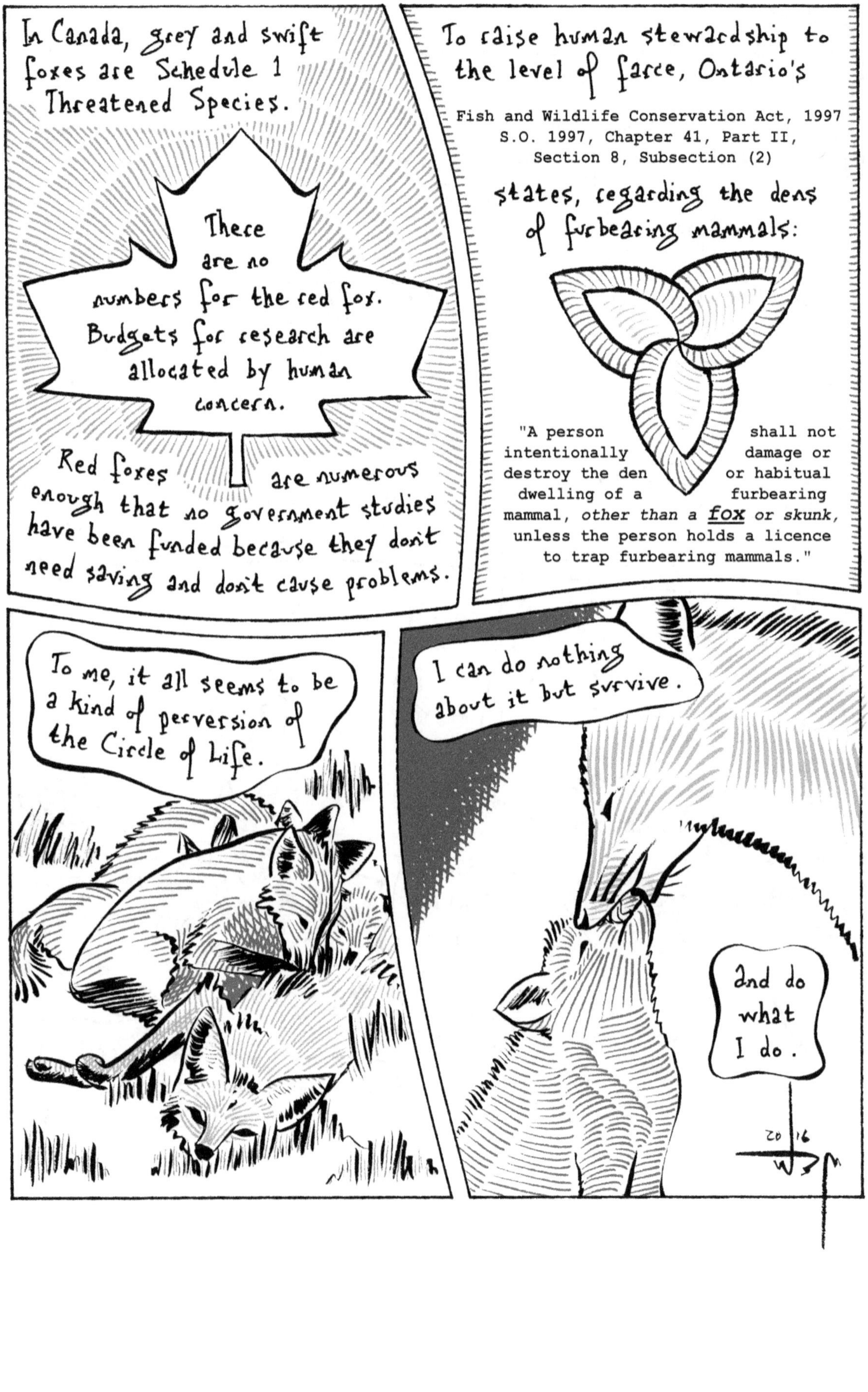
In Canada, grey and swift foxes are Schedule 1 Threatened Species.
There are no numbers for the red fox. Budgets for research are allocated by human concern.
Red foxes are numerous enough that no government studies have been funded because they don't need saving and don't cause problems.
To raise human stewardship to the level of farce, Ontario's
Fish and Wildlife Conservation Act, 1997 S.O. 1997, Chapter 41, Part II, Section 8, Subsection (2)
states, regarding the dens of furbearing mammals:
"A person shall not intentionally damage or destroy the den or habitual dwelling of a furbearing mammal, *other than a* ***fox*** *or skunk*, unless the person holds a licence to trap furbearing mammals."
To me, it all seems to be a kind of perversion of the Circle of Life.
I can do nothing about it but survive.
And do what I do.
2016

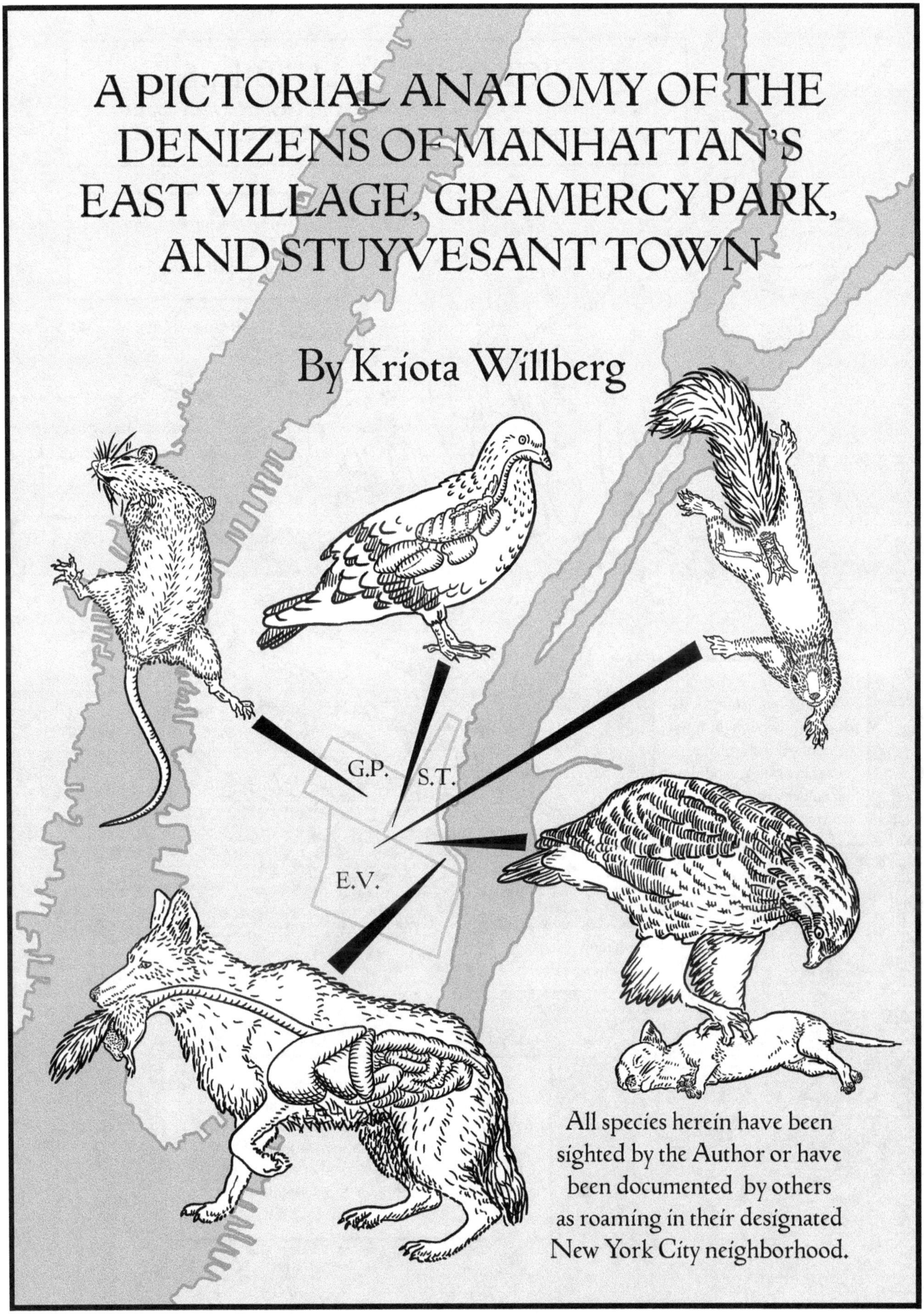
A PICTORIAL ANATOMY OF THE DENIZENS OF MANHATTAN'S EAST VILLAGE, GRAMERCY PARK, AND STUYVESANT TOWN
By Kriota Willberg
G.P.
S.T.
E.V.
All species herein have been sighted by the Author or have been documented by others as roaming in their designated New York City neighborhood.

PIGEONS ON CITIBIKES
The respiratory system of the feral pigeon (*Columba livia domestica*)
at 1st Avenue and 15th Street
The air sacs associated with the lungs of pigeons serve as organs of respiration and thermoregulation.
Some of the air sacs enter into bones, making them lighter.
Air sacs are for cooling and lung ventilation. They direct air through the lungs in one direction only, instead of the in-and-out arrangement of mammal respiratory systems.
Pigeon lungs exchange gases. Air sacs do not.
To keep cool, pigeons pant. The evaporation of moisture in the air sacs helps to keep them cool. They're just like dogs and coyotes, except that dogs rely on evaporation from the mucus membranes of the mouth and upper respiratory system since they don't have air sacs, or feathers, or beaks, or - never mind.
citibike
KW '16

A BROWN RAT GETTING EXERCISE
External reproductive anatomy of the female brown rat (*Rattus novegicus*)
somewhere in the mid-20th streets east of Park Avenue
This is a female rat. She has a clitoris. The size of the clitoris is similar to that of the protruding portion of the male rat penis. Due to the similarity in size it might be difficult to determine the sex of the rat, visually.
Mammary papillae are the external openings for the mammary glands.
She has three pairs of mammary papillae at the thorax.
And three pairs of mammary papillae are located at the abdomen.
This mammary grouping would make it possible for a rat of any gender to wear a modified bikini, displaying midriff fur and retaining modesty.
WARNING! Rats should not attempt to pierce their nipples due to their small size and regular use (see below).
Identification tip: The distance between the rat clitoris and anus is approximately half the distance of the penis to the anus.
As female rats can bear up to twelve litters per year and are likely nursing the pups, they have to be very comfortable with their bodies if they are going to wear bikinis.
Why is determing the sex of a rat important? It isn't unless you are studying rat reproductive organs.
KW 16

SNACK TIME FOR RED-TAILED HAWKS
A comparison of *Buteo jamaicensis* talons and *Canis lupus familiaris* paws on air conditioners in Stuyvesant Town on the Avenue A Loop
These hawks have come up to Stuyvesant Town from Tompkins Square Park for a little snack.*
*Predation of pets by red-tailed hawks is rare but possible.
Red-tailed hawks and chihuahuas are both digitigrade animals, meaning they walk on what are considered "finger" and/or "toe" bones.
Hawks use their talons to constrict and immobilize prey to prevent escape. The hawk will stand on larger prey to hold it still and begin to eat it alive! Death is usually the result of exsanguination or organ failure.
Chihuahuas have only rudimentary coordination where gripping prey is concerned.*
*This statement is not supported by research.
4
1
3
2
3
2
4
1
5
1
4
3
2
3 4
2
5
1
KW '16

MELANISTIC EASTERN GRAY SQUIRREL AND MINIATURE BALLERINA ENJOY THEIR AFTERNOON
A comparison of lower extremity positioning between *Sciurus carolinensis* and *Homo sapiens* (*minimus*) on a tree at 14th Street at 1st Avenue
The hips and lower extremities of squirrels and humans share structural similarities.
FOOT: Squirrels easily twist their feet into a position called inversion.
FOOT: Ballet dancers avoid sickling (a.k.a. inversion) of their feet at all costs.
Humans are bipedal and plantigrade locomotors. That is they walk on two feet and contact the ground with the full foot (the heel through toes).
KNEE: This joint can be described as a modified hinge joint.
HIP: The positions* of this ballerina's hip creates what is known in dance as "turn out."
*This hip combines the positions of abduction (sideways movement) and lateral rotation (twisting the hip away from the dancer's midline).
KNEE: Flexion (bending) is roughly similar for both species.
HIP: Squirrels have excellent "turn out" without having to warm up or stretch.
When ballerinas dance *en pointe*, they temporarily become ungulants, using the tips of the toes to sustain body weight while in motion.
Squirrels are digitigrade locomotors.
The lack of sharp claws at the ends of her fingers and toes will cause this miniature ballerina to slip off the tree trunk at any moment.
KW 16

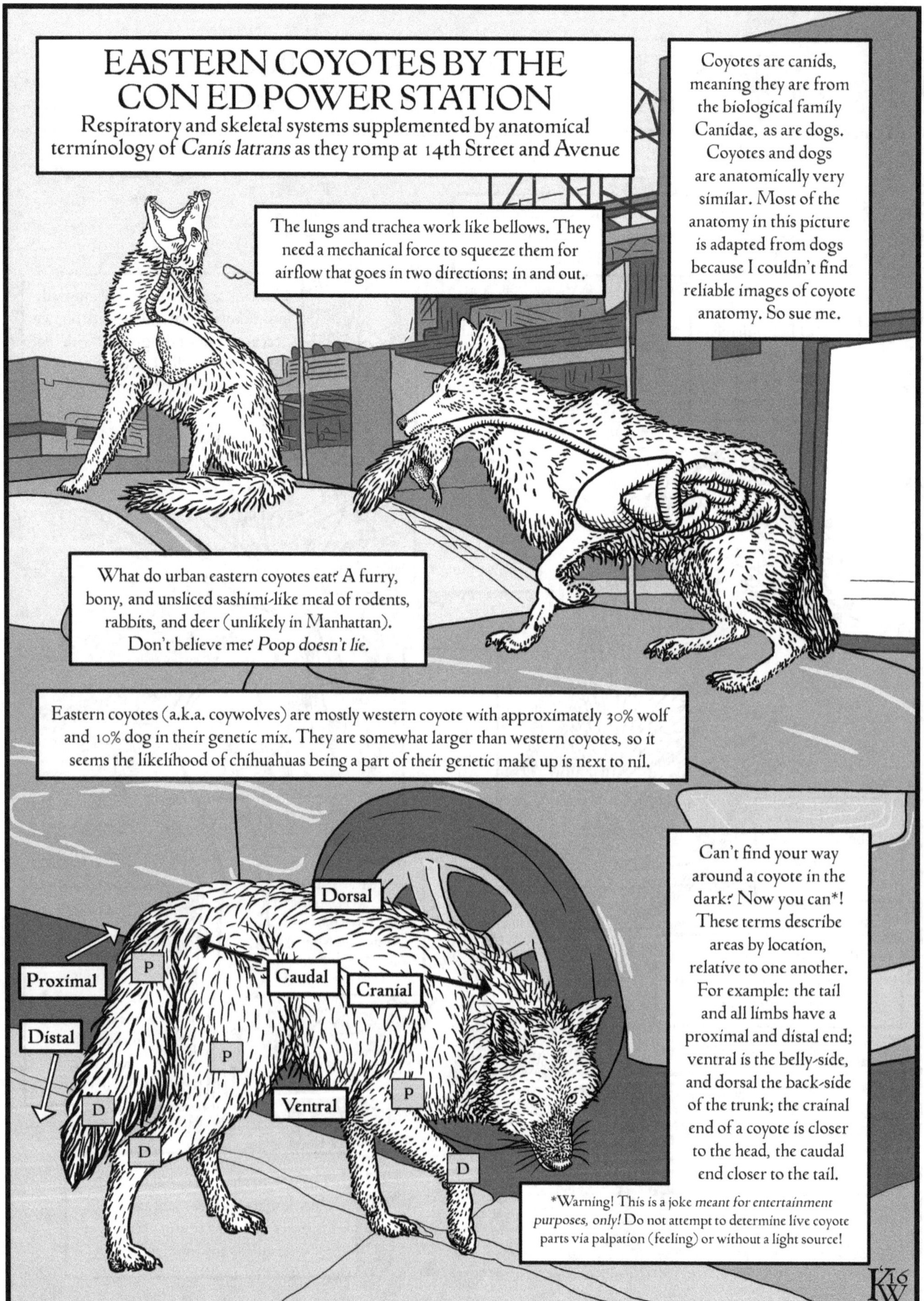
EASTERN COYOTES BY THE CON ED POWER STATION
Respiratory and skeletal systems supplemented by anatomical terminology of *Canis latrans* as they romp at 14th Street and Avenue
Coyotes are canids, meaning they are from the biological family Canidae, as are dogs. Coyotes and dogs are anatomically very similar. Most of the anatomy in this picture is adapted from dogs because I couldn't find reliable images of coyote anatomy. So sue me.
The lungs and trachea work like bellows. They need a mechanical force to squeeze them for airflow that goes in two directions: in and out.
What do urban eastern coyotes eat? A furry, bony, and unsliced sashimi-like meal of rodents, rabbits, and deer (unlikely in Manhattan). Don't believe me? *Poop doesn't lie.*
Eastern coyotes (a.k.a. coywolves) are mostly western coyote with approximately 30% wolf and 10% dog in their genetic mix. They are somewhat larger than western coyotes, so it seems the likelihood of chihuahuas being a part of their genetic make up is next to nil.
Dorsal
Proximal
P
Caudal
Cranial
Distal
P
P
D
Ventral
D
D
Can't find your way around a coyote in the dark? Now you can*! These terms describe areas by location, relative to one another. For example: the tail and all limbs have a proximal and distal end; ventral is the belly-side, and dorsal the back-side of the trunk; the crainal end of a coyote is closer to the head, the caudal end closer to the tail.
*Warning! This is a joke *meant for entertainment purposes, only!* Do not attempt to determine live coyote parts via palpation (feeling) or without a light source!

"Preserving The Silurian Age" by Kes Bradley & David Kirkham

SQUEEEEEZE!!!
GOTCHA!
TOSS!
CHOMP!
AS THE AGES HURTLED FORWARD, THESE CREATURES WERE LEFT FROZEN IN TIME UNDER THE SEABED...
THE FOSSILIZATION* PROCESS
* Pure fossilization is actually very rare. It requires perfect conditions for sediments to successfully preserve a body!
(JUST KIDDING!)
R.I.P
IT TAKES MILLIONS OF YEARS BEFORE A FOSSIL FORMS SO IT CAN BE DISCOVERED!

IN THE CENTURIES THAT FOLLOWED, MAN BEGAN TO MINE FOR RICHES.
BY JOVE!
WHAT THEY FOUND WAS RICHER STILL...
SINCE THE INDUSTRIAL REVOLUTION, OVER 700 FOSSILIZED SPECIES HAVE BEEN UNCOVERED IN DUDLEY ALONE!
The township relished these discoveries, so much that in 1866, a trilobite was added to their coat of arms. It became known worldwide as
THE DUDLEY BUG!
TODAY, YOU CAN FIND MANY SPECIMENS OF SILURIAN FOSSILS AT DUDLEY MUSEUM AND ART GALLERY.
CLIK!

MEANWHILE, THE SITE OF THESE DISCOVERIES HAS BECOME A PROTECTED GEO-PARK IN THE WREN'S NEST AREA, WHERE PLENTY OF WONDERS FROM HISTORY ARE WAITING TO BE UNEARTHED!
NO HAMMERS
PUBLIC INFORMATION
WREN'S NEST
EST. 1956
NATURE RESERVE
THAT'S ALL FOR NOW!
© D.K. 2016

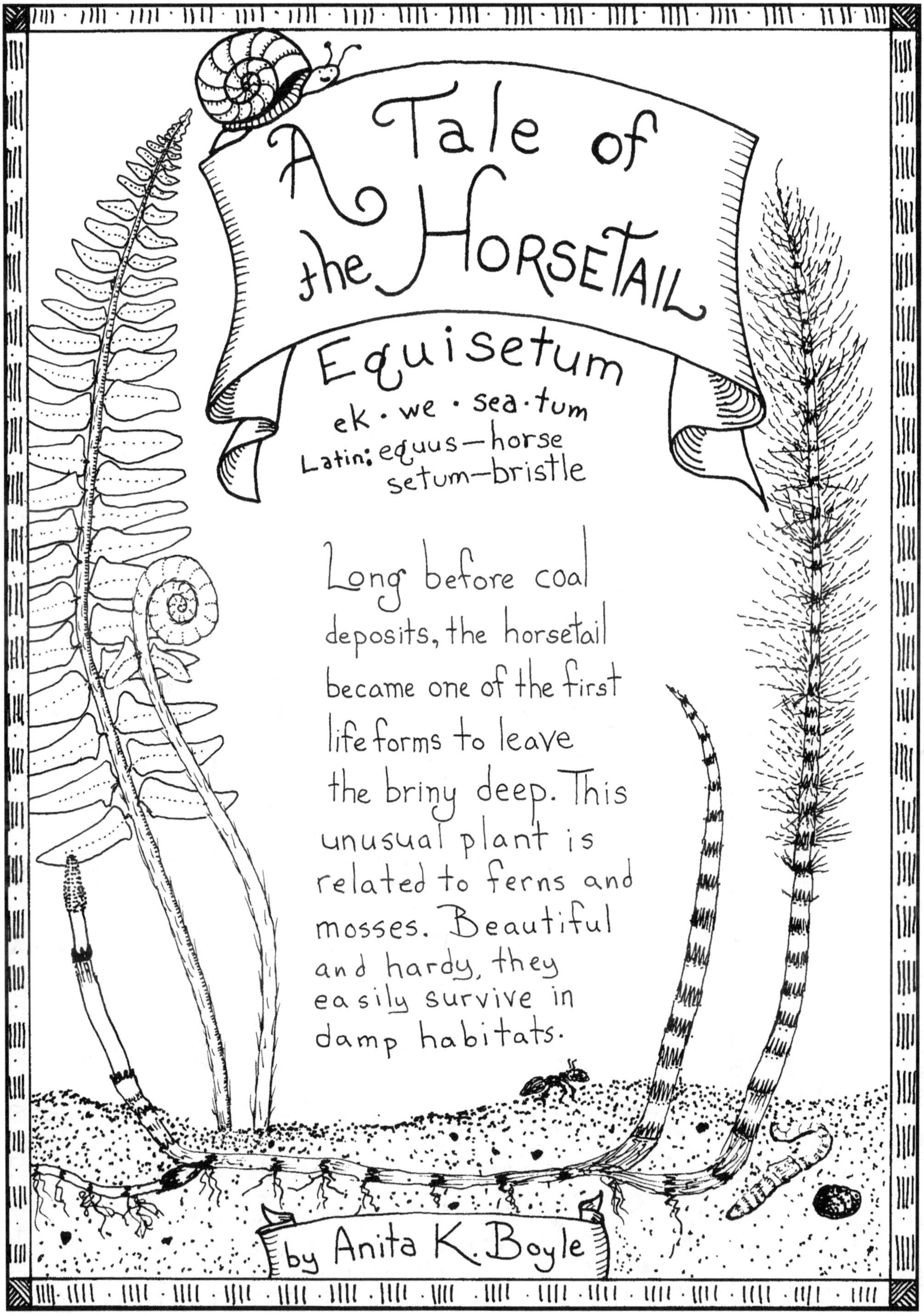
A Tale of the Horsetail
Equisetum
ek · we · sea · tum
Latin: equus—horse
setum—bristle
Long before coal deposits, the horsetail became one of the first life forms to leave the briny deep. This unusual plant is related to ferns and mosses. Beautiful and hardy, they easily survive in damp habitats.
by Anita K. Boyle

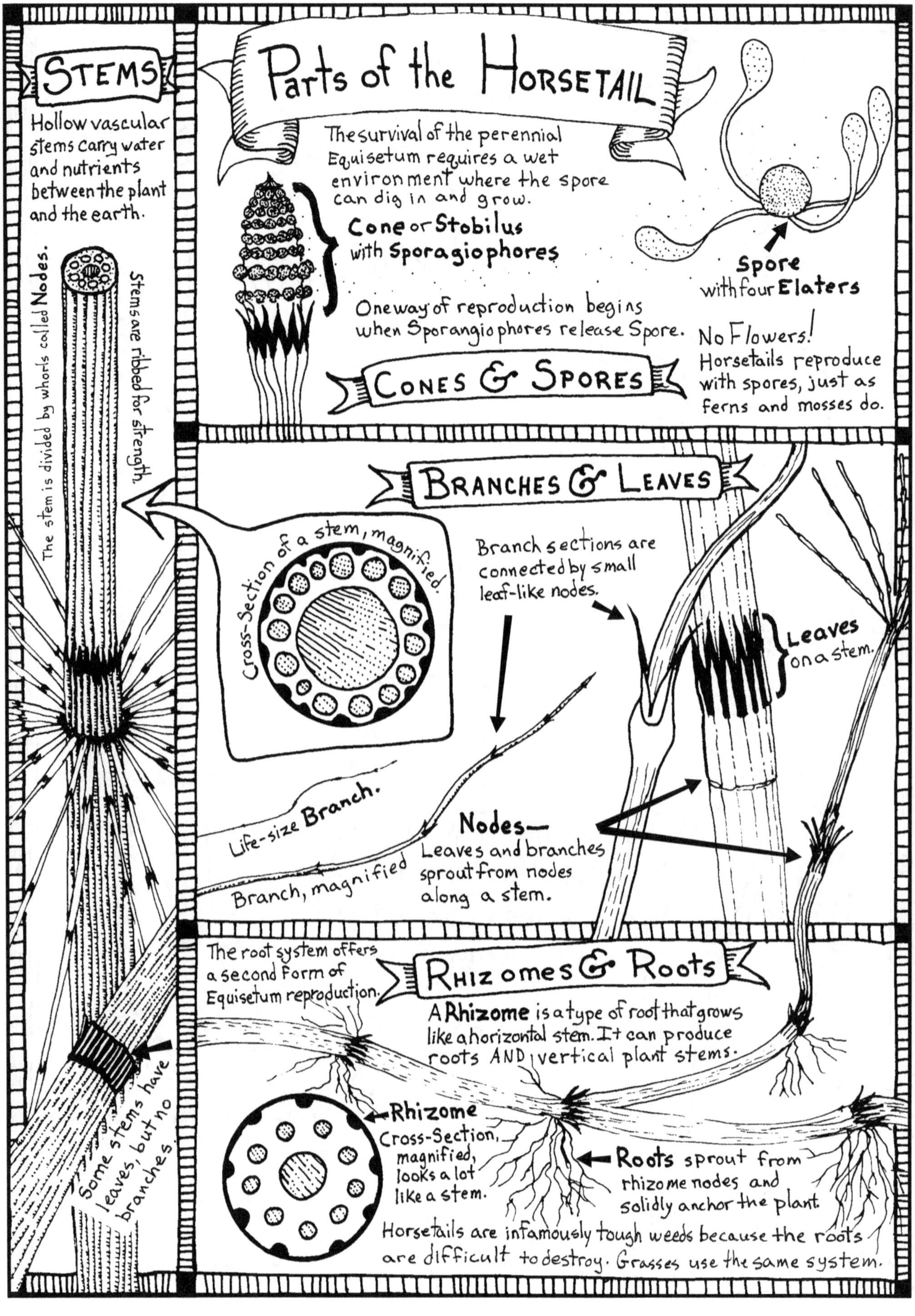
Parts of the Horsetail
Stems
Hollow vascular stems carry water and nutrients between the plant and the earth.
The stem is divided by whorls called Nodes.
Stems are ribbed for strength.
Some stems have leaves, but no branches.
The survival of the perennial Equisetum requires a wet environment where the spore can dig in and grow.
Cone or Stobilus with Sporagiophores
One way of reproduction begins when Sporangiophores release Spore.
Spore with four Elaters
No Flowers! Horsetails reproduce with spores, just as ferns and mosses do.
Cones & Spores
Branches & Leaves
Cross-Section of a stem, magnified.
Branch sections are connected by small leaf-like nodes.
Leaves on a stem.
Life-size Branch.
Branch, magnified
Nodes— Leaves and branches sprout from nodes along a stem.
Rhizomes & Roots
The root system offers a second form of Equisetum reproduction.
A Rhizome is a type of root that grows like a horizontal stem. It can produce roots AND vertical plant stems.
Rhizome Cross-Section, magnified, looks a lot like a stem.
Roots sprout from rhizome nodes and solidly anchor the plant.
Horsetails are infamously tough weeds because the roots are difficult to destroy. Grasses use the same system.

Equisetum telmateia
Genus Latin Translation: from the marshes
Common Name: Giant Horsetail
The Giant Horsetail is the tallest non-tropical horsetail in the world. It resembles its prehistoric ancestors from 350 million years ago, except in height. Back then, one species grew to be 100 feet tall and ate tricerotops for dinner. Well, okay, no, it didn't. It was eaten by dinosaurs itself.
Having survived at least two mass extinctions, it is much hardier than dinosaurs. Proof? Do you see any dinosaurs?
Even the lowly slug has survived the dinosaur eras. In fact, the slug has been around almost twice as long as horsetails have—600 million years. Next time you come upon a slug, show a little respect for once.
The Giant Horsetail is shaped like a chimney brush.
Since the telmateia grows 5-8 feet tall, only a young one would fit on these pages. This tall "Horse-Bristle" prospers in damp, shady places, near springs and seepage lines in the woods.

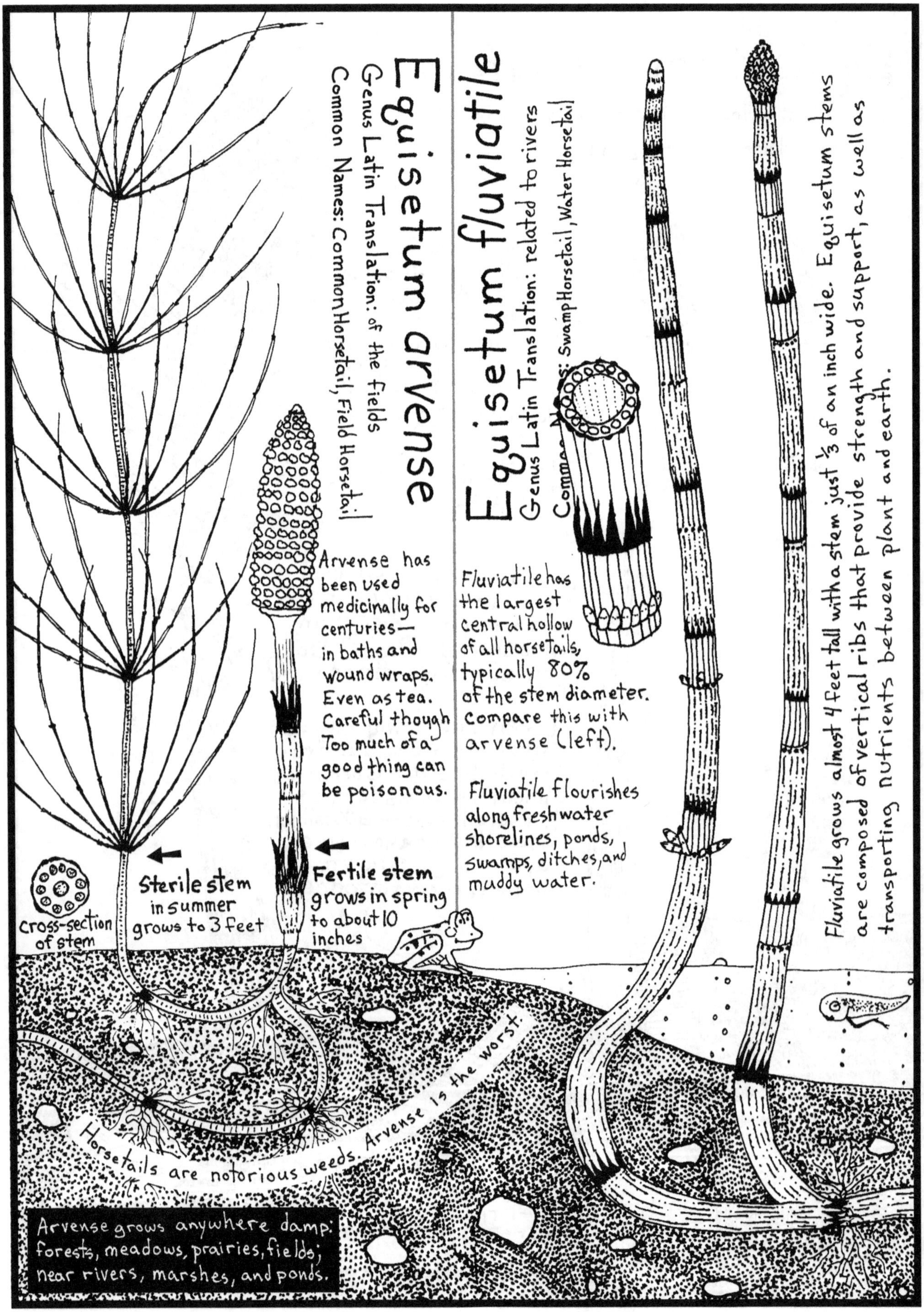
Equisetum arvense
Genus Latin Translation: of the fields
Common Names: Common Horsetail, Field Horsetail
Arvense has been used medicinally for centuries— in baths and wound wraps. Even as tea. Careful though Too much of a good thing can be poisonous.
Sterile stem in summer grows to 3 feet
Fertile stem grows in spring to about 10 inches
cross-section of stem
Horsetails are notorious weeds. Arvense is the worst.
Arvense grows anywhere damp: forests, meadows, prairies, fields; near rivers, marshes, and ponds.
Equisetum fluviatile
Genus Latin Translation: related to rivers
SwampHorsetail, Water Horsetail
Fluviatile has the largest central hollow of all horsetails, typically 80% of the stem diameter. Compare this with arvense (left).
Fluviatile flourishes along freshwater shorelines, ponds, swamps, ditches, and muddy water.
Fluviatile grows almost 4 feet tall with a stem just 1/3 of an inch wide. Equisetum stems are composed of vertical ribs that provide strength and support, as well as transporting nutrients between plant and earth.

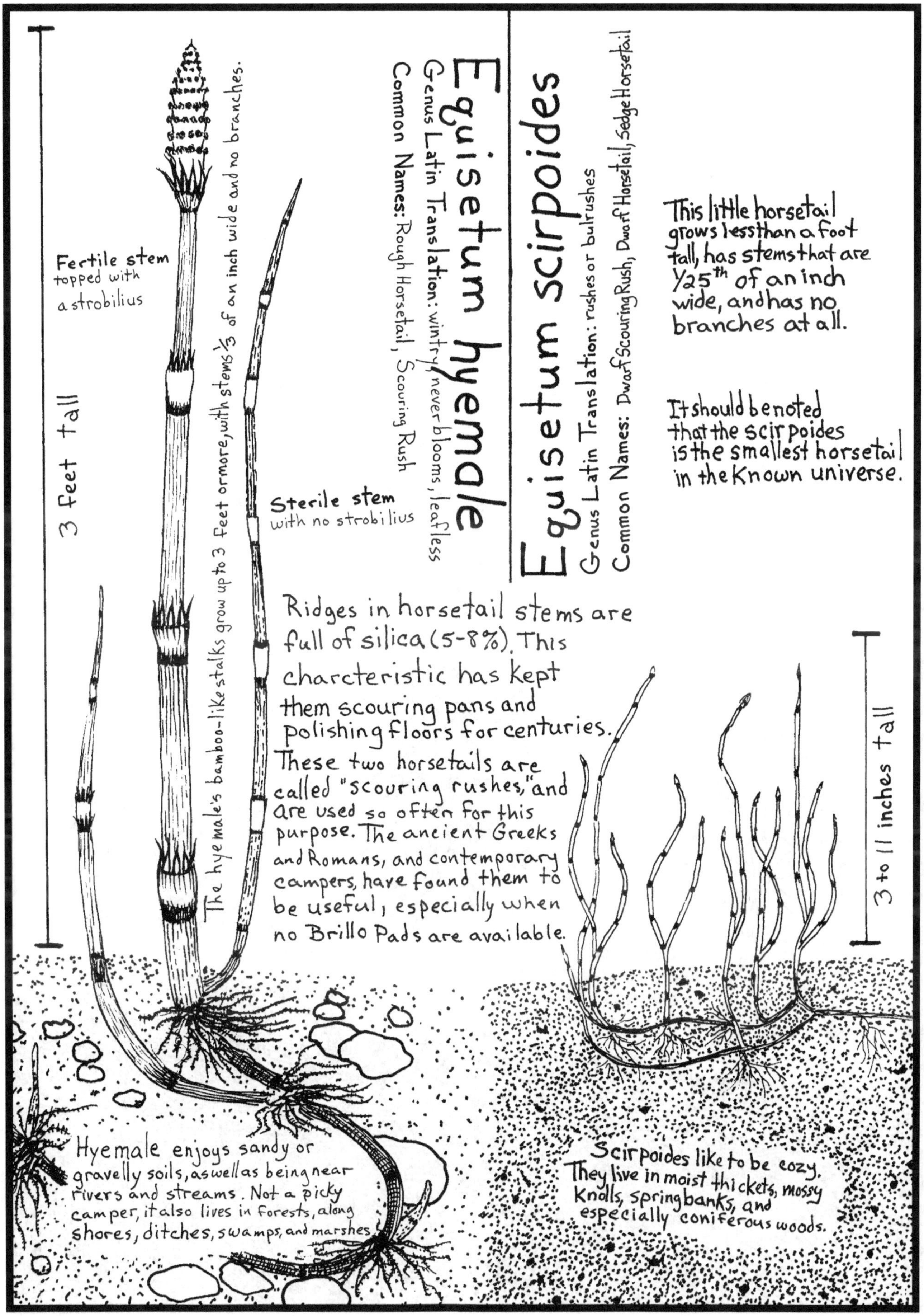
Equisetum hyemale
Genus Latin Translation: wintry, never blooms, leafless
Common Names: Rough Horsetail, Scouring Rush
Equisetum scirpoides
Genus Latin Translation: rushes or bulrushes
Common Names: Dwarf Scouring Rush, Dwarf Horsetail, Sedge Horsetail
This little horsetail grows less than a foot tall, has stems that are 1/25th of an inch wide, and has no branches at all.
It should be noted that the scirpoides is the smallest horsetail in the known universe.
3 feet tall
Fertile stem topped with a strobilius
The hyemale's bamboo-like stalks grow up to 3 feet or more, with stems 1/3 of an inch wide and no branches.
Sterile stem with no strobilius
Ridges in horsetail stems are full of silica (5-8%). This charcteristic has kept them scouring pans and polishing floors for centuries.
These two horsetails are called "scouring rushes," and are used so often for this purpose. The ancient Greeks and Romans, and contemporary campers, have found them to be useful, especially when no Brillo Pads are available.
3 to 11 inches tall
Hyemale enjoys sandy or gravelly soils, as well as being near rivers and streams. Not a picky camper, it also lives in forests, along shores, ditches, swamps, and marshes.
Scirpoides like to be cozy. They live in moist thickets, mossy knolls, spring banks, and especially coniferous woods.

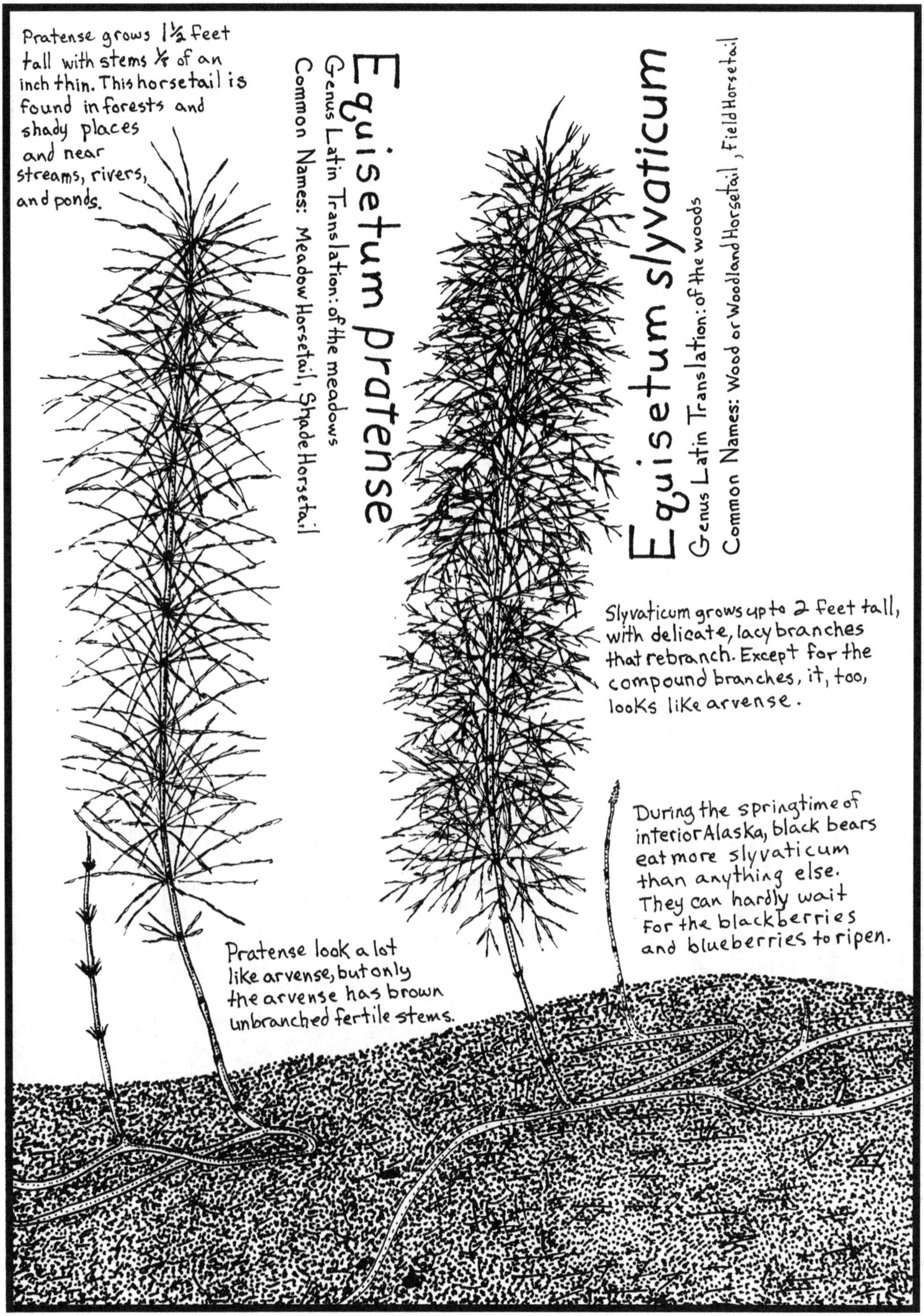
Pratense grows 1½ feet tall with stems ⅛ of an inch thin. This horsetail is found in forests and shady places and near streams, rivers, and ponds.
Equisetum pratense
Genus Latin Translation: of the meadows
Common Names: Meadow Horsetail, Shade Horsetail
Equisetum slyvaticum
Genus Latin Translation: of the woods
Common Names: Wood or Woodland Horsetail, Field Horsetail
Slyvaticum grows up to 2 feet tall, with delicate, lacy branches that rebranch. Except for the compound branches, it, too, looks like arvense.
During the springtime of interior Alaska, black bears eat more slyvaticum than anything else. They can hardly wait for the blackberries and blueberries to ripen.
Pratense look a lot like arvense, but only the arvense has brown unbranched fertile stems.

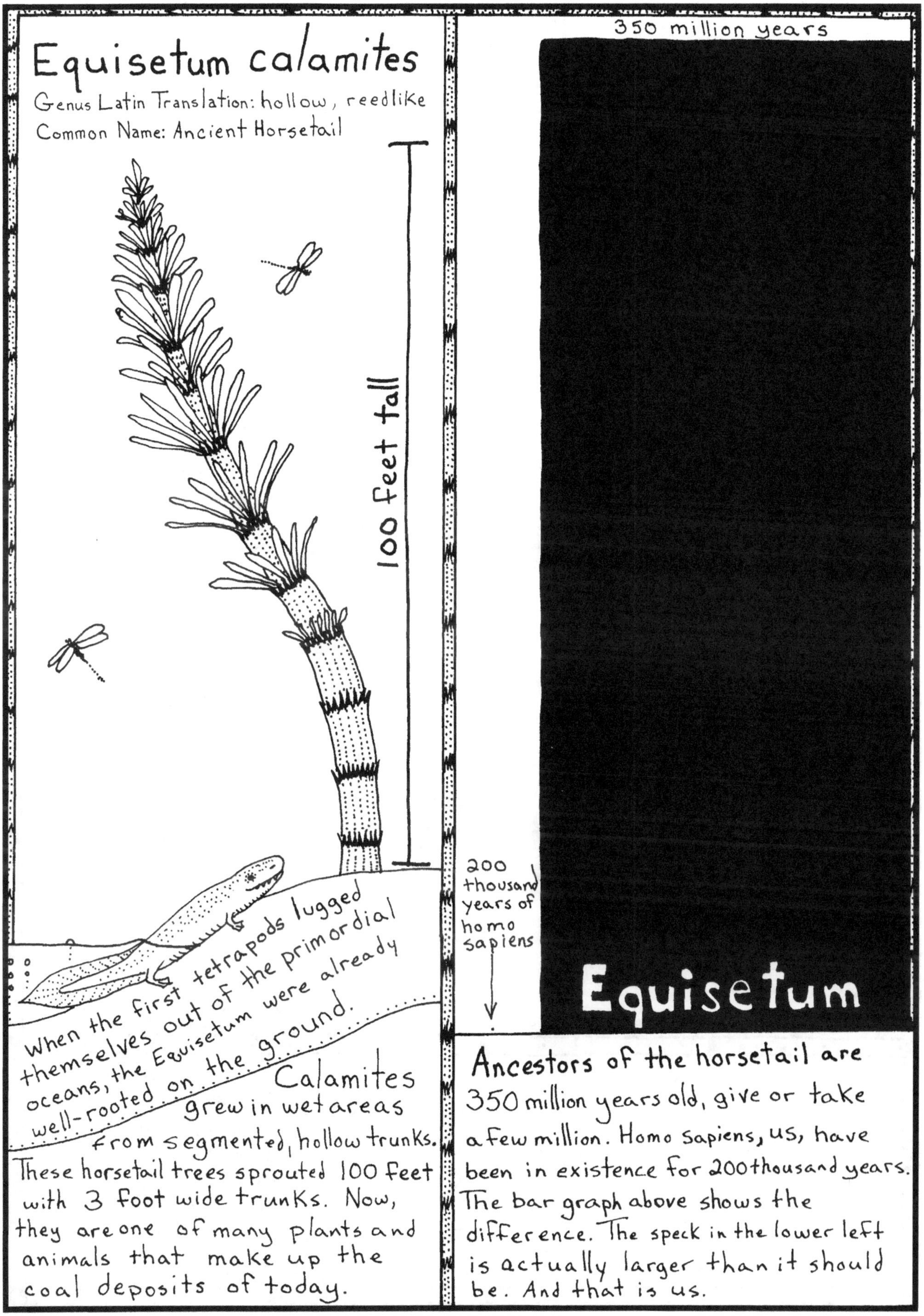
Equisetum calamites
Genus Latin Translation: hollow, reedlike
Common Name: Ancient Horsetail
100 feet tall
When the first tetrapods lugged themselves out of the primordial oceans, the Equisetum were already well-rooted on the ground.
Calamites grew in wet areas from segmented, hollow trunks. These horsetail trees sprouted 100 feet with 3 foot wide trunks. Now, they are one of many plants and animals that make up the coal deposits of today.
350 million years
200 thousand years of homo sapiens
Equisetum
Ancestors of the horsetail are 350 million years old, give or take a few million. Homo sapiens, us, have been in existence for 200 thousand years. The bar graph above shows the difference. The speck in the lower left is actually larger than it should be. And that is us.

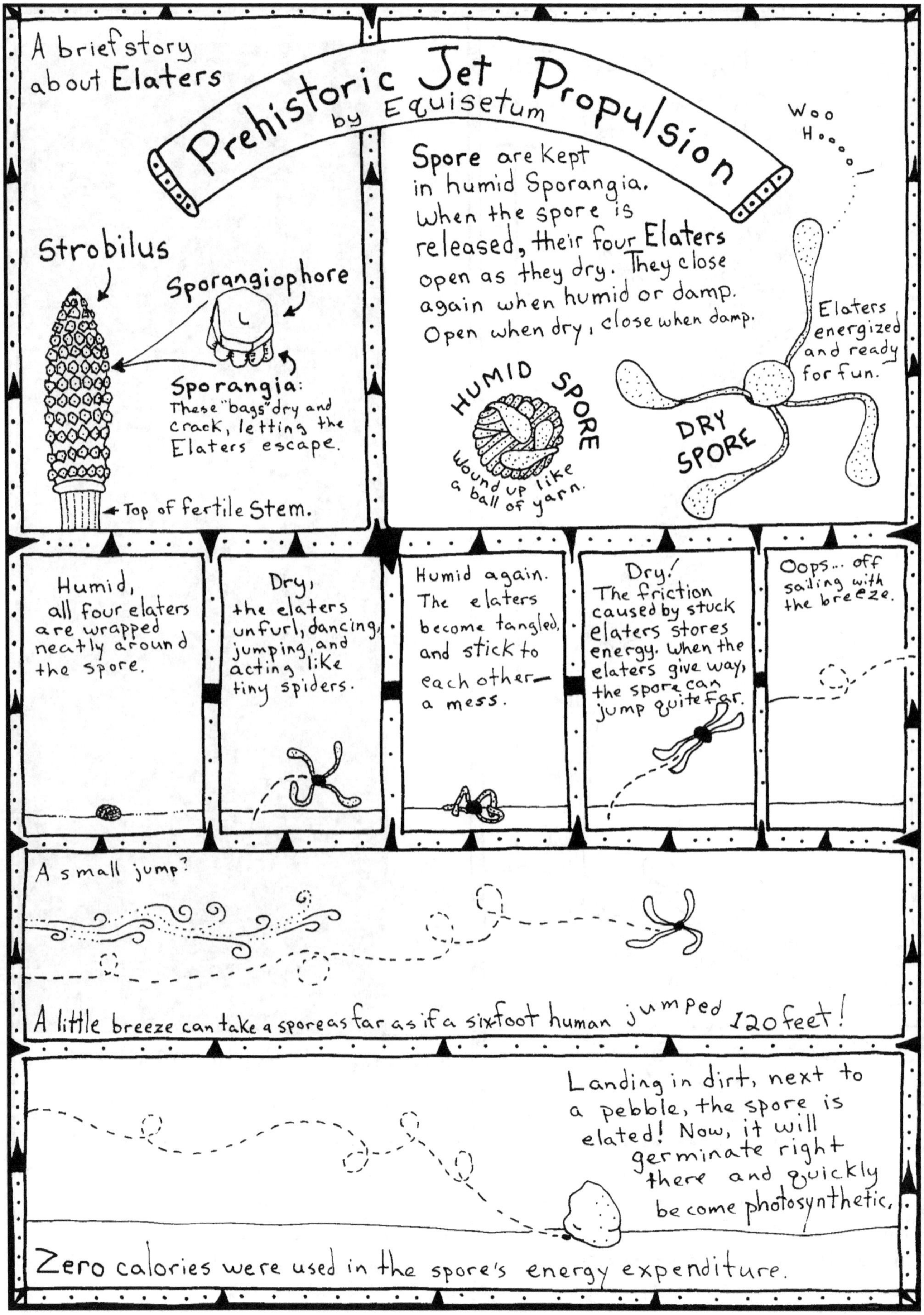
A brief story about Elaters
Prehistoric Jet Propulsion
by Equisetum
Strobilus
Sporangiophore
Sporangia:
These "bags" dry and crack, letting the Elaters escape.
Top of fertile Stem.
Spore are Kept in humid Sporangia. When the spore is released, their four Elaters open as they dry. They close again when humid or damp. Open when dry, close when damp.
Woo Hoooo
Elaters energized and ready for fun.
HUMID SPORE
Wound up like a ball of yarn.
DRY SPORE
Humid, all four elaters are wrapped neatly around the spore.
Dry, the elaters unfurl, dancing, jumping, and acting like tiny spiders.
Humid again. The elaters become tangled, and stick to each other— a mess.
Dry! The friction caused by stuck elaters stores energy. When the elaters give way, the spore can jump quite far.
Oops... off sailing with the breeze.
A small jump?
A little breeze can take a spore as far as if a six foot human jumped 120 feet!
Landing in dirt, next to a pebble, the spore is elated! Now, it will germinate right there and quickly become photosynthetic.
Zero calories were used in the spore's energy expenditure.

Let's meet Joaninha, the **Iberian Lynx**

Lynx pardinus (Temminck, 1827)
The most vulnerable felid in the world
Critically endangered since 2002

By Bernardo O. Dias

The last decrease was when a number of diseases hit the lynx's main source of food.
In 2001, there were less than 100 lynxes left.
Save the Lynx
People took notice. Breeding centers and protected areas were made.
While lynxes are in captivity, biologists and veterinarians watch them through cameras 24/7 to make sure they are healthy and safe. Human contact is kept to a minimum...
...so that Joaninha may learn to hunt and survive in the wild.
Joaninha will grow up, be given a special collar...

... and be released.
The collar allows biologists to find her so that they can check up on her.
It is dangerous out there.
But with any luck, Joaninha will settle down and start a family.
Slowly the lynx is recovering. It is now Endangered since 2015.
Just the beginning, not THE END.

ANGELA BOYLE

ILLUSTRATIONS

Illustration	Illustrator(s)
Animal Rafting	Spratty
Aquariums	Bonesteel
Armadillo	David Kirkham Kim Sucher Iris Yan Ross Wood Studlar
Bat	Kim Sucher
Cicada	Hedj
Coelacanth	Moss Bastille
Cordyceps	Hedj Bernardo O. Dias
Corn	Anita K. Boyle
Crow	Bridget Comeau
Cryptozoology	Melissa McGee
Deep Sea Creatures	Bridget Comeau Simina Popescu
Degus	Allison Bannister
Dodo Bird	Anita K. Boyle
Elephants	Kevin Kite
Ergot	Payden McRoberts
Fillial Cannibalism	Wm Brian MacLean
Flightless Birds	Payden McRoberts
Fossilized Trilobite	Bernardo O. Dias Angela Boyle
Frog	Sophie Argetsinger
Giraffe	Catherine Garbarino
Grasshopper Mouse	Megan Archer Anita K. Boyle

Great White Shark	Kriota Willberg Iris Yan
Great Blue Heron	Sophie Argetsinger
Horsetail	Angela Boyle Lauren Norby
Iberian Lynx	Patricia Maldonado Simina Popescu Sophie Argetsinger
Invasive Plants	Kit Fraser
Jellyfish	Shashwat Mishra Catherine Garbarino
Megafauna	Megan Archer
Moose Brainworm	Melissa McGee
Plainswood Cotton	Kit Fraser
Rattlesnake	David Kirkham
Raven	Kami Koyamatsu
Salmon	Kelly Chen
Seahorse	Kelly Fernandez Kelly Ricker Kelly Chen
Sexual Selection	Spratty
Shorebirds	Melissa McGee
Sperm Whale	Moss Bastille Tom O'Brien
Starlings	Laura Marie Madden
Tardigrade	Vicky Yuh Jerel Dye
Varroa Mite	Alyssa Suzumura Jerel Dye
Vulture	Kelly Swann Shashwat Mishra

Spratty

Animal Rafting

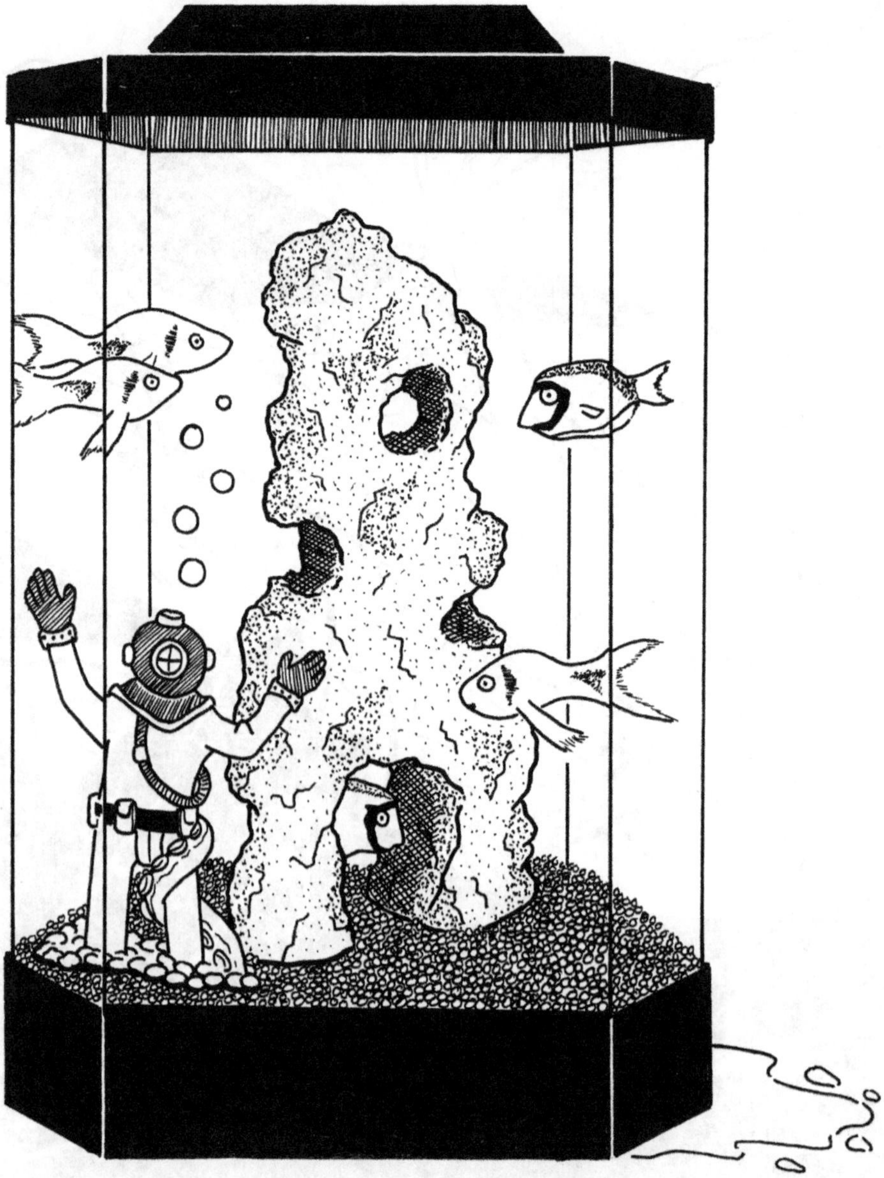

A HISTORY OF AQUARIUMS

Bonesteel

DAVID KIRKHAM

ARMADILLO

Kim Sucher — Armadillo

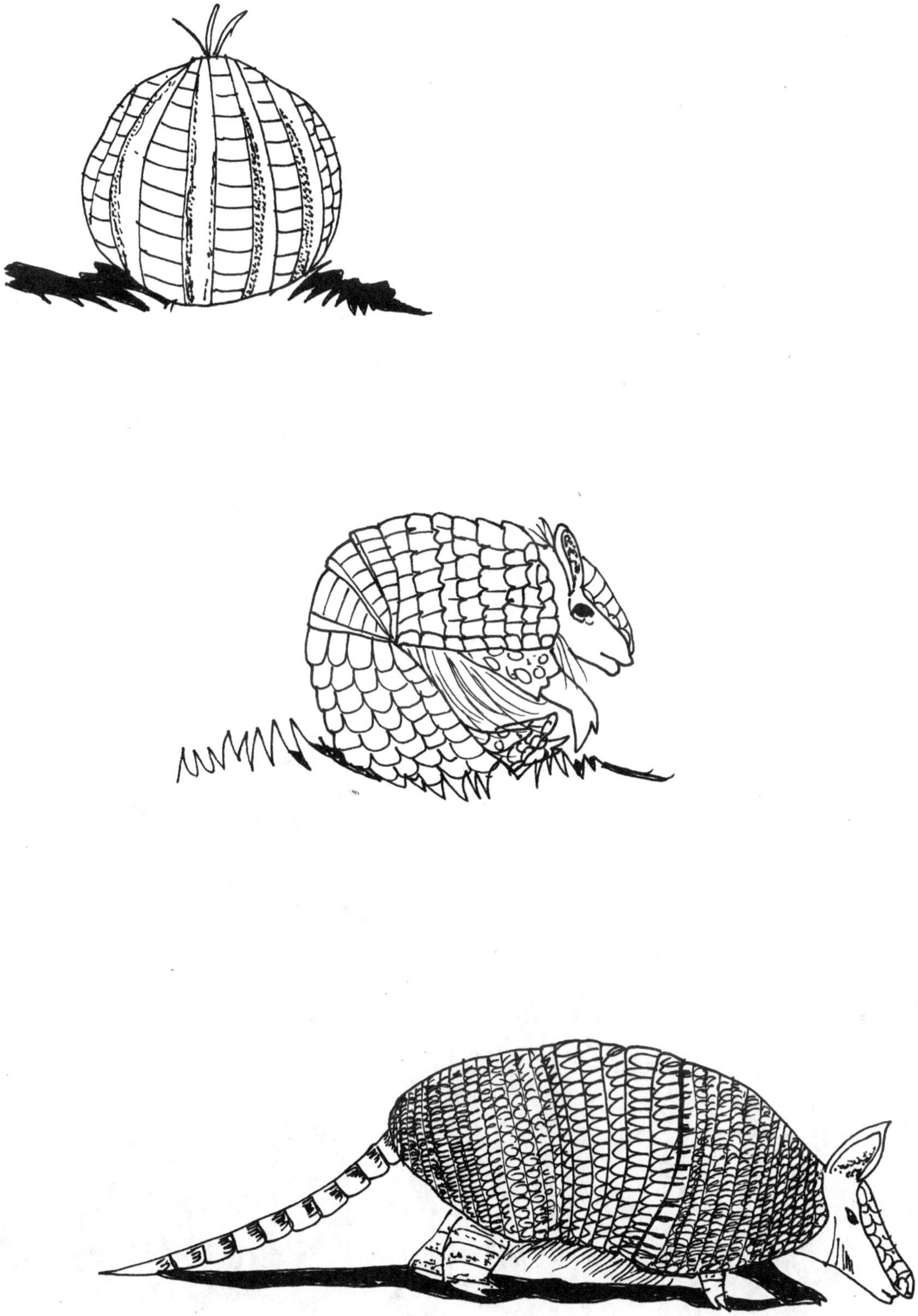

IRIS YAN
ARMADILLO

Ross Wood Studlar

Armadillo

Kim Sucher

Bat

Hedj

Cicada

Moss Bastille

Coelacanth

Hedj

Corcyceps

BERNARDO O. DIAS

CORDYCEPS

Anita K. Boyle

Corn

Bridget Comeau — Crow

Melissa McGee

Cryptozoology

Bridget Comeau

Simina Popescu

Deep Sea Angler Fish

ALLISON BANNISTER

DEGUS'

Anita K. Boyle

THE SOCIAL LIVES OF ELEPHANTS

Kevin Kite

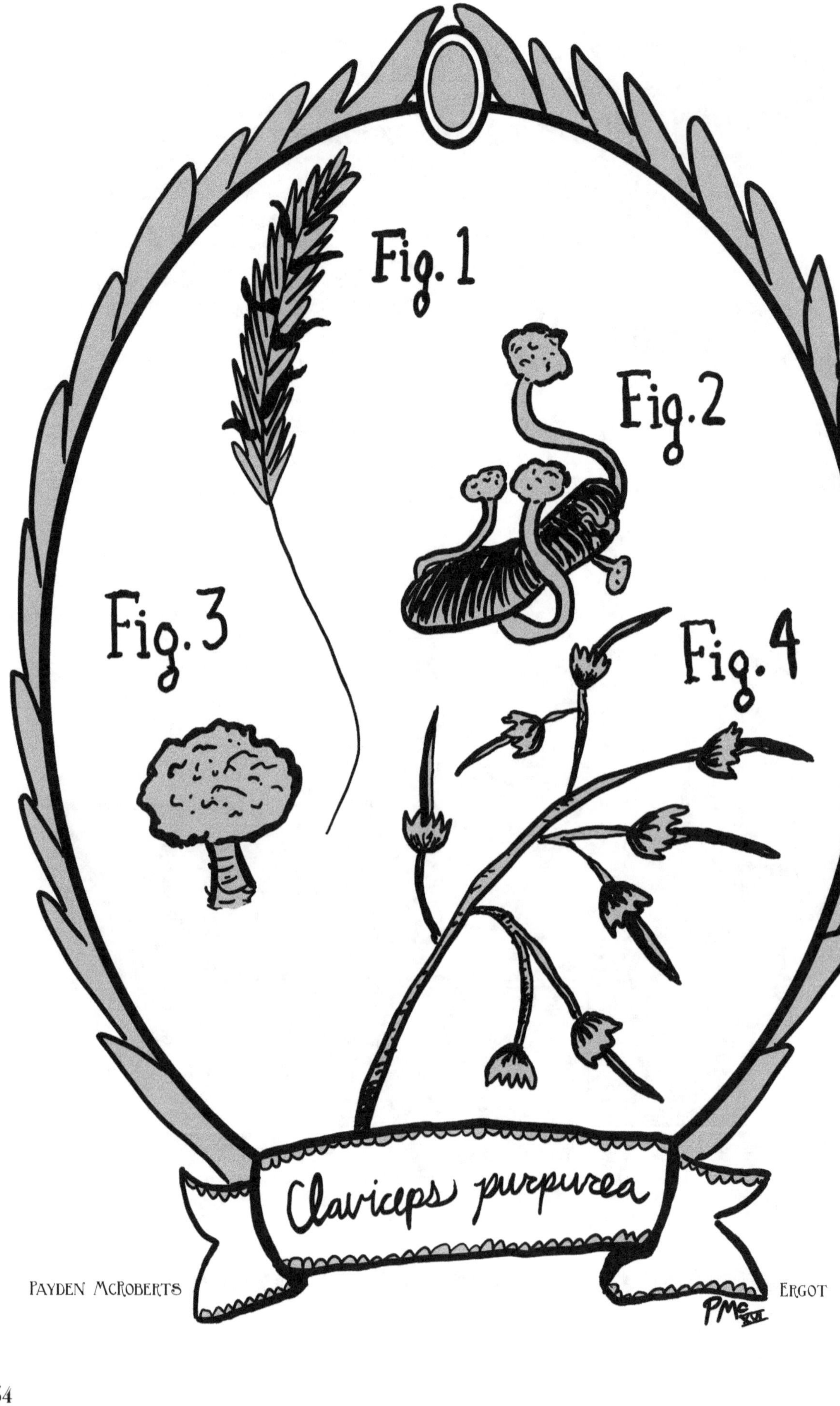

Payden McRoberts

Ergot

Wm Brian MacLean

Payden McRoberts

Flightless Birds

Bernardo O. Dias

Fossilized Trilobite

Angela Boyle

Fossilized Trilobite

SOPHIE ARGETSINGER FROG

Catherine Garbarino — Giraffe

MEGAN ARCHER GRASSHOPPER MOUSE

Anita K. Boyle

Grasshopper Mouse

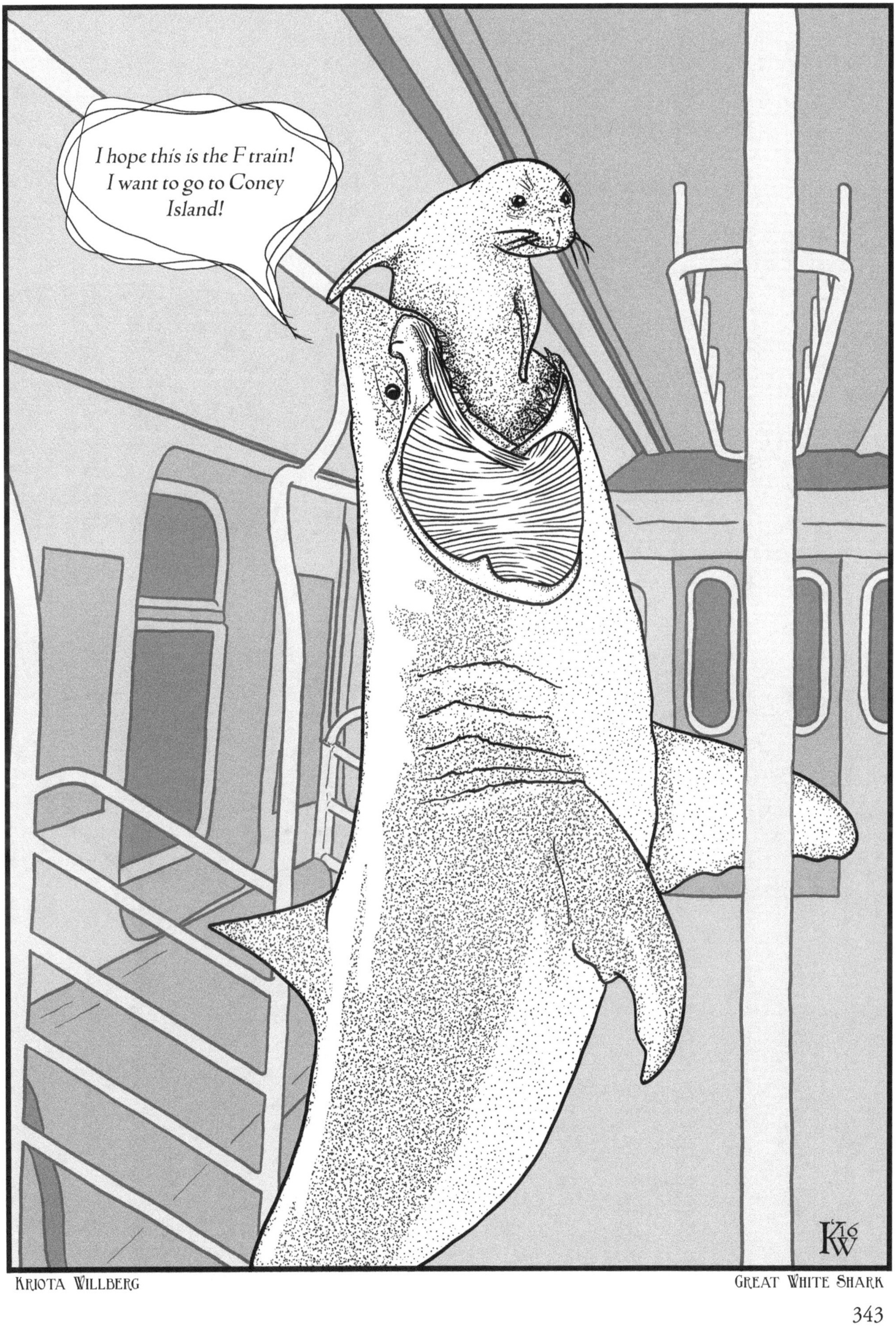

KRIOTA WILLBERG

GREAT WHITE SHARK

Iris Yan
Great White Shark

SOPHIE ARGETSINGER

GREAT BLUE HERON

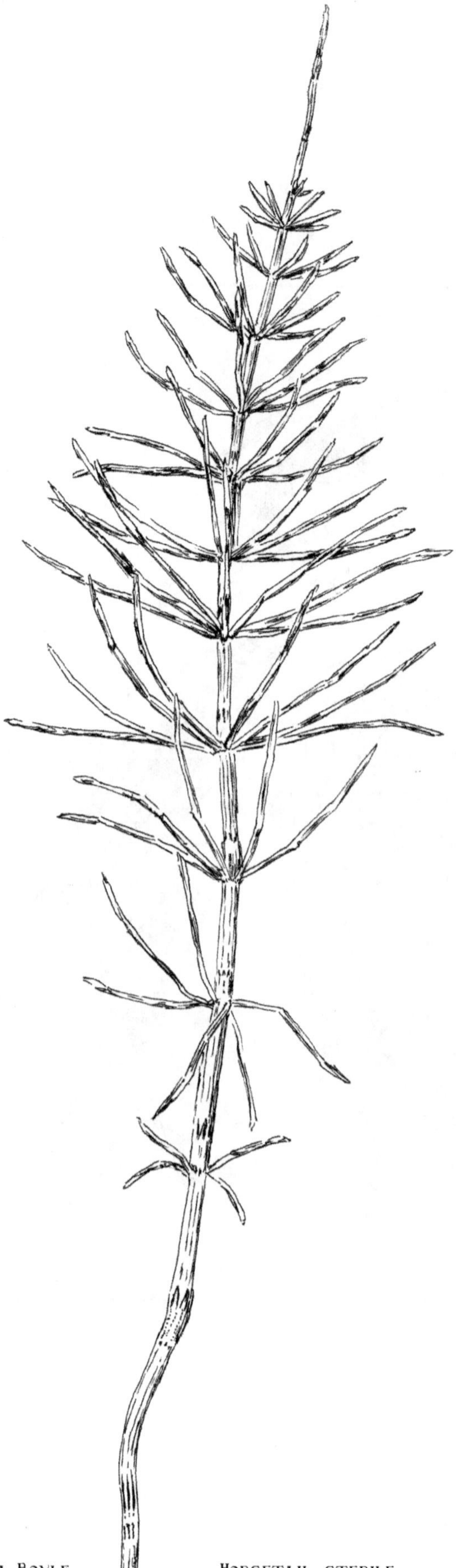

Angela Boyle   Horsetail, sterile

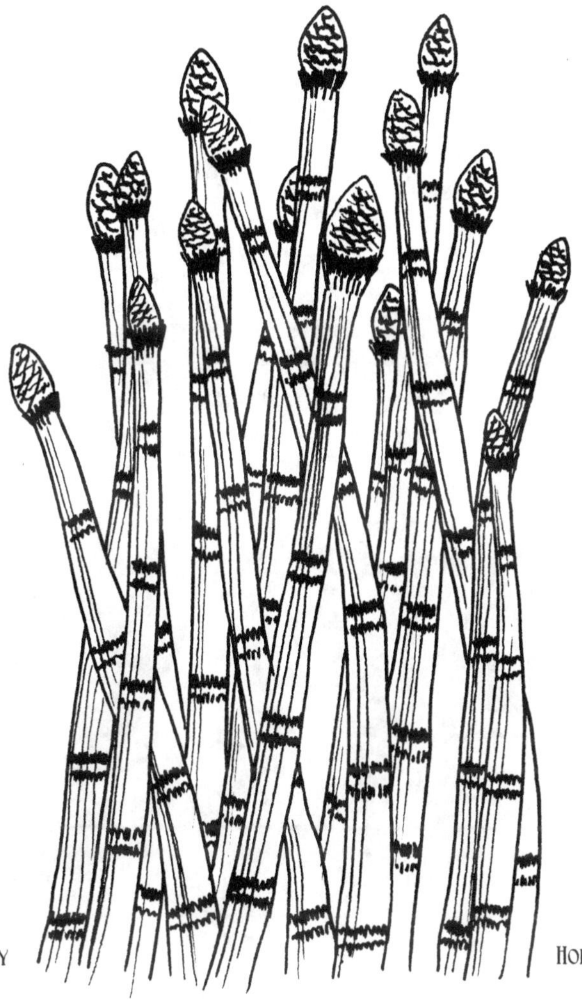

Lauren Norby

Horestail, fertile

Simina Popescu

Iberian Lynx

Sophie Argetsinger

Iberian Lynx

Patricia Maldonado

Iberian Lynx

Kit Fraser — Invasive Plants

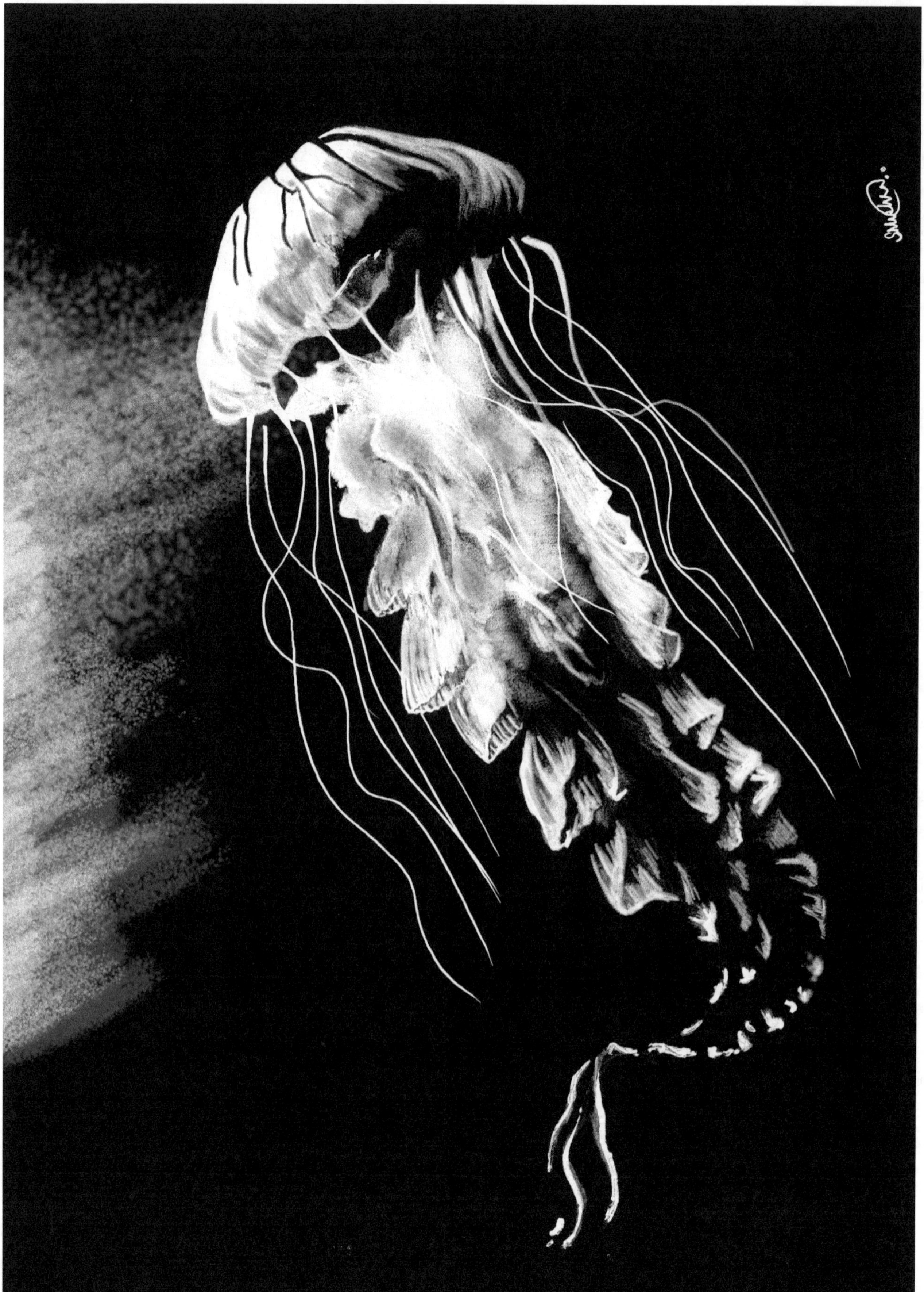

Shashwat Mishra

Jellyfish

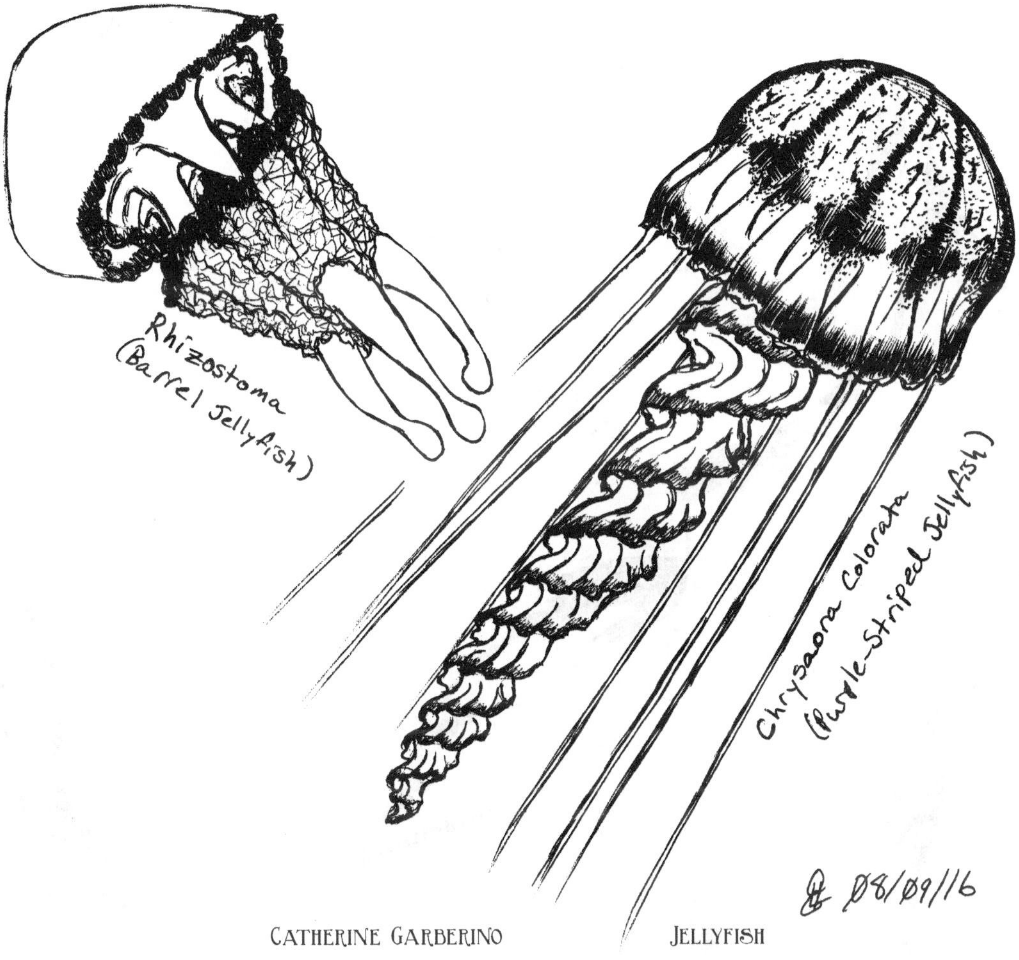

Catherine Garberino Jellyfish

Megan Archer

Megafauna

Melissa McGee

Moose Brainworm

Kit Fraser

Plains Cottonwood

David Kirkham

Rattlesnake

Kami Koyamatsu

Raven

KELLY CHEN SALMON

Kelly Fernandez

Seahorse

KELLY RICKER

SEAHORSE

Kelly Chen

Seahorse

SPRATTY — SEXUAL SELECTION

Melissa McGee

Shorebirds

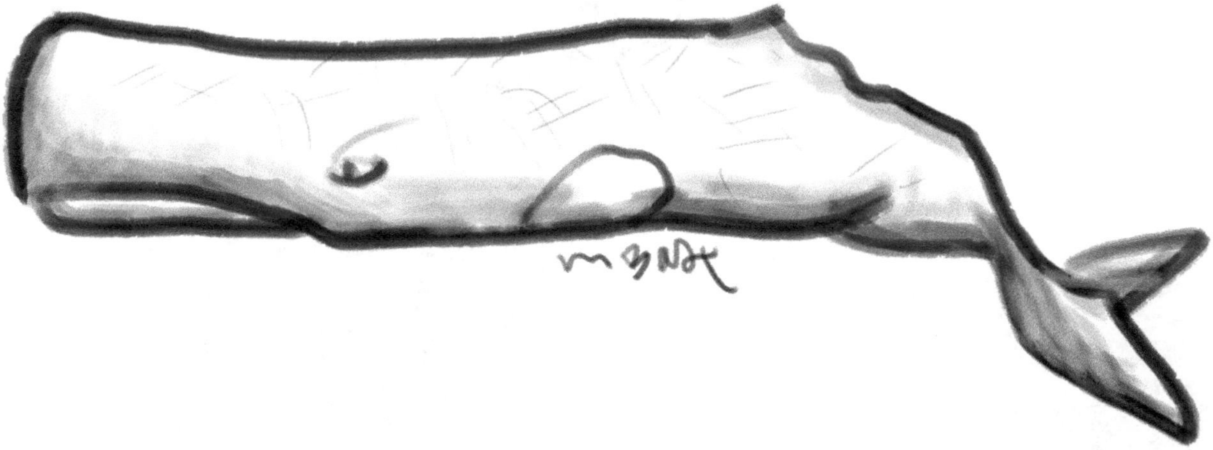

MOSS BASTILLE

SPERM WHALE

Tom O'Brien

Sperm Whale

Laura Marie Madden

Starlings

Vicky Yuh

Tardigrade

JEREL DYE TARDIGRADE

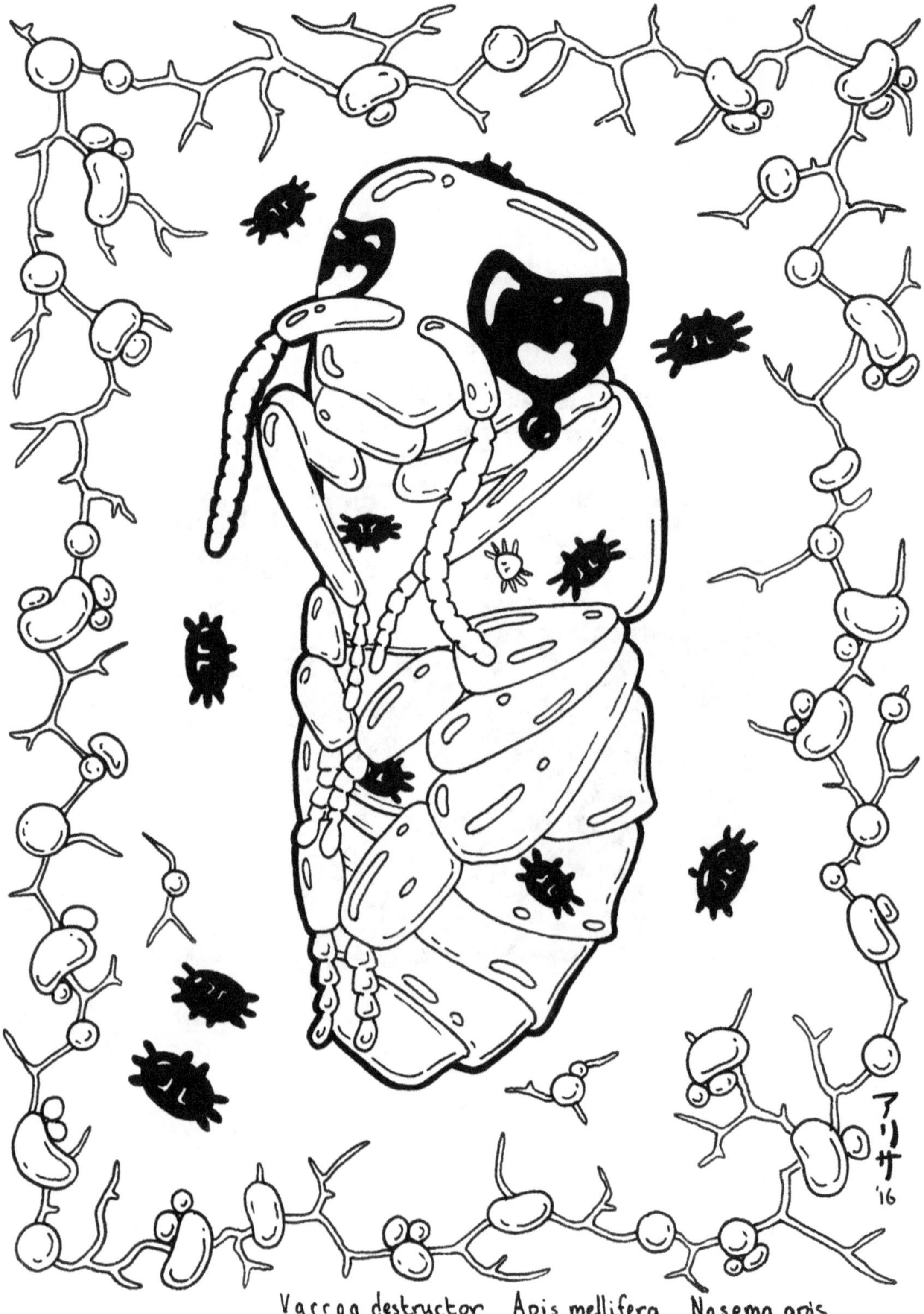

ALYSSA SUZUMUSA

JEREL DYE

VARROA MITES

Kelly Swann

Vulture

Shashwat Mishra Vulture

BIBLIOGRAPHIES

Apis Mellifera and *Varroa Destructor*
by William Scavone

13 Attack Bee 010.MTS. Video, 0:08. Accessed July 9, 2016. https://drive.google.com/file/d/0B_CdCaK7ZPMWQXhLaldhb2g0OE0/view.

Bonney, Richard E. *Hive Management: A Seasonal Guide for Beekeepers*. Charlotte, VT: Garden Way Publishing, 1990.

Conrad, Ross. *Natural Beekeeping: Organic Approaches to Modern Apiculture*. White River Junction, VT: Chelsea Green Publishing, 2007.

Huang, Zachary. "Varroa Mite Reproductive Biology." *EXtension*. July 26, 2013. http://articles.extension.org/pages/65450/varroa-mite-reproductive-biology.

Morse, Roger A. *The New Complete Guide to Beekeeping*. Woodstock, VT: The Countryman Press, 1994.

O'Hanlon, Dan. *Raising Northern Queens*. Vermont Beekeeper's Association 2016 Summer Meeting, Hartland, VT. Attended July 9, 2016.

Oliver, Randy. "Varroa Management." *Scientific Beekeeping*. Updated January 15, 2016. Accessed July 26, 2016. http://scientificbeekeeping.com/varroa-management.

Sanford, Malcom T. and Richard E. Bonney. *Storey's Guide to Keeping Honey Bees*. North Adams, MA: Storey Publishing, 2010.

Seeley, Thomas D. *Honeybee Democracy*. Princeton, NJ: Princeto University Press, 2010.

Wikipedia. "Deformed Wing Virus." *Wikipedia*. Last modified May 29, 2016. https://en.wikipedia.org/wiki/Deformed_wing_virus.

Don't Tread on Us by Spratty

Amarello, Melissa. "A Day in the Life of Rattlesnake Families Advocates for Snake Preservation." *Advocates for Snake Preservation*. October 29, 2011. https://www.snakes.ngo/a-day-in-the-life-of-a-rattlesnake-family.

Butler, Joseph A., Todd W. Hull, and Richard Franz. "Neonate Aggregations and Maternal Attendance of Young in the Eastern Diamondback Rattlesnake, *Crotalus adamanteus*." *American Society of Ichthyologists and Herpetologists (ASIH)* 1995, no, 1 (February 15, 1995): 196–98. doi:10.2307/1446814.

Clark, Rulon W., William S. Brown, Randy Stechert, and Harry W. Greene. "Cryptic Sociality in Rattlesnakes (*Crotalus horridus*) Detected by Kinship Analysis." *Biology Letters* 8, no. 4 (August 23, 2012): 523–25. doi:10.1098/rsbl.2011.1217.

Henderson, Doug, and Dennis Paulson. "Snakes of North America." *University of Pittsburgh*. Accessed September 15, 2016. http://www.pitt.edu/~mcs2/herp/SoNA.html.

Milius, Susan. "The Social Lives of Snakes: From Loner to Attentive Parent." *Science News* 165, no. 13 (March 27, 2004): 200. http://www.phschool.com/science/science_news/articles/social_lives_snakes.html.

National Wildlife Federation. "Rattlesnake." Accessed October 3, 2016. https://www.nwf.org/Wildlife/Wildlife-Library/Amphibians-Reptiles-and-Fish/Rattlesnakes.aspx.

Pachniewska, Amanda. "Interview with Snake Behavior Researcher Melissa Amarello." *Animal Cognition*. January 12, 2016. http://www.animalcognition.org/2016/01/12/interview-with-snake-behavior-researcher-melissa-amarello.

Holy Fire by Moss Bastille

Encyclopedia Britannica. "Saint Anthony of Egypt." *Encyclopaedia Britannica*. Accessed June 3, 2016. https://www.britannica.com/biography/Saint-Anthony-of-Egypt.

Schumann, Gail L. "Ergot of Rye." *The Plant Health Instructor*. Last updated 2005. doi:10.1094/PHI-I-2000-1016-01.

Stewart, Amy and Briony Morrow-Cribbs. *Wicked Plants: The Weed that Killed Lincoln's Mother and Other Botanical Atrocities*. Chapel Hill, NC: Algonquin Books of Chapel Hill, 2009.

Thomson, M. "Pont-Saint-Esprit Poisoning: Did the CIA Spread LSD?" *BBC News*. August 23, 2010. http://www.bbc.com/news/world-10996838.

Wong, G. J. "Ergot of Rye - I: Introduction and History." University of Hawai'i at Mānoa. Accessed June 3, 2016. http://www.botany.hawaii.edu/faculty/wong/BOT135/LECT12.HTM.

Armadillos by Kate Lacour

Ferro, Shaunacy. "8 Absurd Facts About Armadillos." *Mental Floss Magazine*, October 19, 2015. Accessed May 15, 2016. http://mentalfloss.com/article/69790/8-more-absurd-facts-about-armadillos.

Potts, Steve. *The Armadillo*. Mankato, Minnesota: Captone Press, 1998.

Aphid-Farming Ants by Megan Archer

Chadwick, Steven. "Aphids Rounded up by 'Cowboy' Ants." *Beach Metro*, June 29, 2015. Accessed September 1, 2016. http://www.beachmetro.com/2015/06/29/aphids-rounded-cowboy-ants.

Grant, Bonnie L. "How Do Aphids Help Ants: Controlling Aphids and Ants on Plants." *Gardening Know How*. March 30, 2015. Accessed September 1, 2016. http://www.gardeningknowhow.com/plant-problems/pests/insects/controlling-aphids-and-ants.htm.

Imperial College London. "Herding Aphids: How 'Farmer' Ants Keep Control of Their Food." *ScienceDaily*. October 11, 2007. Accessed September 1, 2016. https://www.sciencedaily.com/releases/2007/10/071009212548.htm.

Don't Call It a Comeback by Hedj

Balon, Eugene K. "Coelacanthiformes (Coelacanths)." *Grzimek's Animal Life Encyclopedia*. Ed. Michael Hutchins, et al. 2nd ed. Vol. 4: Fishes I. Detroit: Gale, 2004. 189–196. Gale Virtual Reference Library. 29 June 2016.

Weinberg, Samantha. *A Fish Caught in Time: The Search for the Coelacanth*. London: Fourth Estate, 2000.

Sepia by Ethan Kocak

Gosse, Philip Henry. *Handbook to the Marine Aquarium*, 2nd Edition Reprint: 1–10. London: Euston Grove Press, 2010.

"Jeanne Villepreux-Power." *Epigenesys*. Accessed August, 2016. http://www.epigenesys.eu/en/science-and-you/women-in-science/682-jeanne-villepreux-power.

"Meet Jeanne Villepreux-Power—The Inventor of the Aquarium." *Women of Science Rock*. March 19, 2014. http://womenrockscience.tumblr.com/post/80117386771/meet-jeanne-villepreux-power-the-inventor-of-the.

Wijgerde, Tim. "Victorian Pioneers of the Marine Aquarium." *Advanced Aquarist*. Last modified February, 2016. http://www.advancedaquarist.com/2016/2/aafeature2.

Wikipedia. "Philip Henry Gosse." *Wikipedia*. Last modified July 5, 2016.https://en.wikipedia.org/wiki/Philip_Henry_Gosse.

———. "Anna Thynne." *Wikipedia*. Last modifed July 2, 2016. https://en.wikipedia.org/wiki/Anna_Thynne.

Resurrection by Ross Wood Studlar

Fox, Douglas. "Cold-Blooded Solutions to Warm-Blooded Problems." *Exploratorium*. Last accessed October 3, 2016. https://www.exploratorium.edu/frogs/woodfrog/index.html.

Frazer, Jennifer. "Ancient Moss Revived After Ages on Ice." *National Geographic*, March 18, 2014. http://news.nationalgeographic.com/news/2014/03/140317-ancient-moss-frozen-life-biology-science.

Jönsson, K. Ingemar, Elke Rabbow, Ralph O. Schill, Mats Harms-Ringdahl, and Petra Rettberg. "Tardigrades Survive Exposure to Space in Low Earth Orbit." *Current Biology* 18, no.17 (2008): 729–731. doi:10.1016/j.cub.2008.06.048.

La Farge, Catherine, Krista H. Williams, and John H. England. "Regeneration of Little Ice Age Bryophytes Emerging from a Polar Glacier with Implications of Totipotency in Extreme Environments." *Proceedings of the National Academy of Sciences* 24, no. 110 (2013): 9839–44. doi:10.1073/pnas.1304199110.

Marchand, Peter J. *Life in the Cold: An Introduction to Winter Ecology*. Hanover: University Press of New England, 1996.

McClanahan, Lon L., Rodolfo Ruibal, and Vaughan H. Shoemaker. "Frogs and Toads in Deserts." *Scientific American*, March 1, 1994. http://faculty.bennington.edu/~sherman/how%20do%20animals%20work%3F/frogs%20and%20toads%20desert.pdf.

Miller, William R. "Tardigrades." *American Scientist* 99, no. 5 (September/October 2011): 384. doi:10.1511/2011.92.384.

Roads, Esme, Royce E. Longton and Peter Convey. "Millennial timescale regeneration in a moss from Antarctica." *Current Biology* 24, no. 6 (2014): 222–223. doi:10.1016/j.cub.2014.01.053.

Seki, Kunihiro and Masato Toyoshima. "Preserving Tardigrades Under Pressure." *Nature* 395 (1998): 853–854. doi:10.1038/27576.

Valentino, Jim and Steve Montano. "The (Secret) Origin of the Guardians of the Galaxy." *Guardians of the Galaxy* 1, no. 26, July 1992. Marvel Comics.

Vitt, Dale H., Barbara Crandall-Stotler, and Andrew Wood. "Bryophytes: Surival in a Dry World through Tolerance and Avoidance." *Plant Ecology and Evolution in Harsh Enviroments*. New York: Nova Science, 2014. http://bryophytes.plant.siu.edu/PDFIles/Vitt%20et%20al.%20chapter.pdf.

"Water Loss by Design." Living Desert Zoo and Gardens State Park, 1504 Miehls Drive N., Carlsbad, NM. August 1, 2013.

The Tireless Tardigrade by Kevin Kite and Michelle McCauley

BBC. "Big Five Mass Extinction Events." *BBC Earth*. Accessed September 30, 2016. Last updated October 2014. http://www.bbc.co.uk/nature/extinction_events.

Fox-Skelly, Jasmin. "Tardigrades Return from the Dead." *BBC Earth*. March 13, 2015. http://www.bbc.com/earth/story/20150313-the-toughest-animals-on-earth.

Glime, J. M. "Tardigrade Survival." *Bryophyte Ecology*. Michigan Technological University and the International Association of Bryologists, 2013. Last updated July 2, 2013. http://bryoecol.mtu.edu. http://www.bryoecol.mtu.edu/chapters_VOL2/5-1Tardigrades_Survival.pdf.

Guidetti, Roberto, Tiziana Altiero, Roberto Bertolani, Pasqualina Grazioso, and Lorena Rebecchi. "Survival of Freezing by Hydrated Tardigrades Inhabiting Terrestrial and Freshwater Habitats." *Zoology* 114, no. 2 (April 2011): 123–128. doi:10.1016/j.zool.2010.11.005.

Ingemar Jönsson, K., Elke Rabbow, Ralph O. Schill, Mats Harms-Ringdahl, and Petra Rettberg. "Tardigrades Survive Exposure to Space in Low Earth Orbit." *Current Biology* 18, no. 17 (September 9, 2008): R729–R731. doi:10.1016/j.cub.2008.06.048.

Ono, F., M. Saigusa, T. Uozumi, Y. Matsushima, H. Ikeda, N.L. Saini, and M. Yamashita. "Effect of High Hydrostatic Pressure on to Life of the Tiny Animal Tardigrade." *Journal of Physics and Chemistry of Solids* 69, no. 9 (September 2008): 2297–2300. doi:10.1016/j.jpcs.2008.04.019.

Rebecchi, Lorena, Tiziana Altiero, Roberto Guidetti, Michele Cesari, Roberto Bertolani, Manuela Negroni, and Angela M. Rizzo. "Tardigrade Resistance to Space Effects: First Results of Experiments on the LIFE-TARSE Mission on FOTON-M3 (September 2007)." *Astrobiology* 9, no. 6 (August 11, 2009): 581–591. doi:10.1089/ast.2008.0305.

Tsujimoto, Megumu, Satoshi Imura, and Hiroshi Kanda. "Recovery and Reproduction of an Antarctic Tardigrade Retrieved from a Moss Sample Frozen for over 30 Years." *Cryobiology* 72, no. 1 (February, 2016): 78–81. doi:10.1016/j.cryobiol.2015.12.003.

Wikipedia. "Extinction Event." *Wikipedia*. Accessed September 30, 2016. https://en.wikipedia.org/wiki/Extinction_event.

Apex by Catherine Garbarino

Discovery. *Sharkopedia*. Last accessed June 18, 2016. http://sharkopedia.discovery.com.

"Diving at Guadalupe." *Dive the World*. Last accessed June 19, 2016. http://www.dive-the-world.com/diving-sites-mexico-guadalupe-island.php.

Graf, Martin. "Science Expedition to Guadalupe Island." *Shark Diver*. Last accessed June 19, 2016. http://www.sharkdiver.com/science-expedition-to-guadalupe-island.

Roy, Katherine. *Neighborhood Sharks: Hunting With the Great Whites of California's Farallon Islands*. New York: David Macaulay Studio, 2015.

Introducing Salmon by Kit Fraser

Greenberg, Paul. "Salmon." *In Four Fish: The Future of the Last Wild Food*, 15–79. New York: Penguin Books, 2010.

Kurlansky, Mark. *World Without Fish*. New York: Workman Publishing Company, 2011.

"The Pebble Proposal." *Save Bristol Bay*. Last accessed October 3, 2016. http://www.savebristolbay.org/pebbleproposal.

"The Salmon's Lifecycle, Its Habitat, Threats and Concerns." *Atlantic Salmon Trust*. Last accessed October 4, 2016.

http://www.atlanticsalmontrust.org/salmon-life-cycle-habitat-threats-and-concerns.html.

Stand Tall by Kelly M. Ricker

Coe, Malcolm J. "'Necking' Behaviour in the Giraffe." *Journal of Zoology* 151, no. 3 (March 1967): 313–21. doi:10.1111/j.1469-7998.1967.tb02117.x.

Estes, Richard. *The Behavior Guide to African Mammals: Including Hoofed Mammals, Carnivores, Primates*, 202–07. University of California Press, 2012.

San Diego Zoo Global. "Mammals: Giraffes." *San Diego Zoo Animals*. Last accessed October 4, 2016. http://animals.sandiegozoo.org/animals/giraffe.

Terrors of the Deep Sea by Shashwat Mishra

Deep Sea Creatures. Last accessed October 4, 2016. http://www.deepseacreatures.org.

Monterey Bay Aquarium Research Institute. *MBARI*. Last accessed October 4, 2016. http://www.mbari.org.

Sea and Sky. Last accessed October 4, 2016. http://www.seasky.org.

World Wide Fund for Nature. *WWF*. Last accessed October 4, 2016. http://wwf.panda.org.

Bats in Flight by Tom O'Brien

"Bat Facts." *Habitat For Bats*. Last accessed August 10, 2016. http://www.habitatforbats.org/?page_id=28.

Dumont, Elizabeth R. "Bone Density and the Lightweight Skeletons of Birds." *Proceedings of the Royal Society B* 227, no. 1691 (March 17, 2010). doi:10.1098/rspb.2010.0117.

Hutchinson, John R. "Vertebrate Flight: Chiropteran Flight." University of California Museum of Paleontology, November 1, 1996. Last accessed August 10, 2016. http://www.ucmp.berkeley.edu/vertebrates/flight/bats.html.

Marden, James H. "Maximum Lift Production During Takeoff in Flying Animals." *Journal of Experimental Biology* 130 (1987): 235–58. Accessed August 10, 2016. http://jeb.biologists.org/content/130/1/235.

Speer, B. R. "Chiroptera: Night Fliers." University of California Museum of Paleontology. August 20, 1995. Last accessed August 10, 2016. http://www.ucmp.berkeley.edu/mammal/eutheria/chiroptera.html.

Raven and Crow by Kelly Fernandez

Audobon. "Identification Pitfalls—Crows and Ravens." *EBird.* March 22, 2011. Last accessed June 28, 2016. http://ebird.org/content/ebird/news/crows_and_ravens.

Boarman, William I., and Bernd Heinrich. "Common Raven (*Corvus corax*)." *Birds of North America Online*, January 1, 1999. Last accessed July 05, 2016. http://bna.birds.cornell.edu/bna/species/476.

Cornell University. "American Crow." *All About Birds*. Last accessed June 27, 2016. https://www.allaboutbirds.org/guide/american_crow/id.

———. "Common Raven." *All About Birds*. Last accessed June 27, 2016. https://www.allaboutbirds.org/guide/Common_Raven/id.

Marzluff, John M., and Tony Angell. *In the Company of Crows and Ravens*. New Haven: Yale University Press, 2005.

McGowan, Kevin. "Similar Species: Crows and Ravens." *All About Birds*. August 03, 2012. Last accessed June 27, 2016. https://www.allaboutbirds.org/similar-species-crowsand-ravens.

Verbeek, N. A., and C. Caffrey. "American Crow (*Corvus brachyrhynchos*)." *Birds of North America Online*. 2002. Last accessed July 5, 2016. http://bna.birds.cornell.edu/bna/species/647/articles/introduction.

Wood, Frances, and Dennis Paulson. "Ravens and Crows—Who Is Who" *BirdNote*. September 2014. Last accessed June 28, 2016. http://birdnote.org/show/ravens-andcrows-who-who.

The Purifiers by Kim Sucher

BBC. "Lammergeier." *BBC Nature*. Last updated October 2014. http://www.bbc.co.uk/nature/life/Bearded_Vulture.

Benson, Elizabeth P. "The Vulture: The Sky and the Earth." *Palenque Round Tables* 10 (1996): 309319. http://www.mesoweb.com/pari/publications/RT10/Vulture.pdf.

BirdLife International. "Andean Condor *Vultur gryphus*." *Data Zone*. Last accessed August 27, 2016. http://www.birdlife.org/datazone/speciesfactsheet.php?id=3822.

Cameron Park Zoological and Botanical Society. "New World Vultures vs. Old World Vultures." *Waco Cameron Park Natural Zoo Habitat*. Last accessed August 27, 2016. http://www.cameronparkzoo.com/wp-content/uploads/2011/08/Vulture-Facts.pdf.

"Cinereous Vulture: (*Aegypius monachus*)." *World Association of Zoos and Aquariums*. Accessed August 26, 2016. http://www.waza.org/en/zoo/visit-the-zoo/birds-of-prey-1254385523/aegypius-monachus.

Cornell University. "Turkey Vulture (*Cathartes aura*)." *All About Birds*. Last accessed August 27, 2016. https://www.allaboutbirds.org/guide/Turkey_Vulture/id.

Cota-Larson, Rhishja. "16 Things You Might Not Know About Vultures." *Annamiticus*. Accessed August 27, 2016. http://annamiticus.com/2012/08/27/16thingsyoumightnotknowaboutvultures.

International Union for Conservation of Nature and Natural Resources. "*Gymnogyps californianus* (California Condor)." The IUCN Red List of Threatened Species. Accessed August 28, 2016. http://www.iucnredlist.org/details/22697636/0.

Sector, Mondo. "A Sky Burial—The Sacred Solemn Funeral Rite of Tibet." Last updated September 10, 2006. http://www.ciolek.com/WWWVLPages/TibPages/tibburial.html.

Stoyanova, Yva, Nikolai Stefanov, and Josef K. Schmutz. "Twig Used as a Tool by the Egyptian Vulture (*Neophron percnopterus*)," *Journal of Raptor Research* 44, no. 2 (June 2010): 154–156. doi:10.3356/JRR-09-20.1.

van Lawick-Goodall, Jane, and Hugo van Lawick-Goodall. "Use of Tools by the Egyptian Vulture, *Neophron percnopterus*," *Nature* 212 (December 24, 1966): 1468–69. doi:10.1038/2121468a0.

Wildscreen. "Eurasian Griffon (*Gyps fulvus*)." Wildscreen Arkive. Accessed August 27, 2016. http://www.arkive.org/eurasian-griffon/gypsfulvus.

Diving in the Deep by Lisa Lowndes & Melissa McGee

Bird, Johnathan. "Sperm Whales: The Deep Divers of the Ocean." *Oceanic Research Group*, June 5, 2007. Accessed April 10, 2016. http://www.oceanicresearch.org/education/wonders/spermwhales.htm.

Boyd, Jade. "Deep-diving Whales Could Hold Answer for Synthetic Blood." *Rice University News and Media*,

September 25, 2015. Accessed April 10, 2016. http://news.rice.edu/2015/09/25/deep-diving-whales-could-hold-answer-for-synthetic-blood-2.

Burnett, Mark. "Inside Nature's Giants: The Sperm Whale." YouTube video, 52:05. Posted by "chocol8milk," August 6, 2014. https://www.youtube.com/watch?v=aXccTHXPYfM.

Gray, Richard. "How the Sperm Whale Can Hold Its Breath for 90 Minutes." *The Telegraph*, June 13, 2013. Accessed April 10, 2016. http://www.telegraph.co.uk/news/science/science-news/10119516/How-the-sperm-whale-can-hold-its-breath-for-90-minutes.html.

"How do Whales Breathe?" *Whale and Dolphin Conservation*. Accessed April 15, 2016. http://uk.whales.org/blog/2012/10/how-do-whales-breathe.

"Sperm Whale Facts." *Whale Facts*. Accessed April 10, 2016. http://www.whalefacts.org/sperm-whale-facts.

Underwater Tails by Kelly Swann

BioExpedition Publishing. *Seahorse Worlds*. Accessed October 4, 2014. http://www.seahorseworlds.com.

Census of Marine Life. "How Many Fish in the Sea? Census of Merine Life Launches First Report." *Science Daily*. October 24, 2003. https://www.sciencedaily.com/releases/2003/10/031024064333.htm.

Danielson, Sentor. "Seahorse Fathers Take Reins in Childbirth." *National Geographic News*. June 14, 2002. http://news.nationalgeographic.com/news/2002/06/0614_seahorse_recov.html.

Science X Network. "Why the Seahorse's Tail Is Square." *Phys.org*. July 2, 2015. http://phys.org/news/2015-07-seahorse-tail-square.html.

Smithsonian National Museum of Natural History. "How Many Seahorse Species?" *Ocean Portal*. Last accessed October 4, 2016. http://ocean.si.edu/ocean-photos/how-many-seahorse-species.

The Seahorse Trust. Accessed October 4, 2016. http://www.theseahorsetrust.org.

Matriarch by Salakjit

Elephants Forever. Last accessed October 5, 2016. http://www.elephantsforever.co.za.

Elephant Voices. Last accessed October 5, 2016. http://www.elephantvoices.org.

"National Geographic Elephant Queen." YouTube video, 43:59. Posted by "ma2017n," October 27, 2015. https://youtu.be/YHD4ouuBmho.

Norris, Paul F. *Animal Wise: A Celebration of Animal Intelligence in Its Many Forms*. Last accessed October 4, 2016. http://www.animalwise.org.

World Wildlife Fund. *WWF*. Last accessed October 5, 2016. http://www.worldwildlife.org.

At Sea by Alyssa Suzumura

Barnes, D. K. "Biodiversity: Invasions by Marine Life on Plastic Debris." *Nature* 416, no. 6883 (April 25, 2002): 808–809. doi:10.1038/416808a.

Lawrence, E. 1998. "Iguanas Ride the Waves." *Nature* (October 15, 1998). doi:10.1038/news981015-3.

Moore, P. D. "Biogeography: Springboards for Springtails." *Nature* 418, no. 6896 (July 25, 2002): 381. doi:10.1038/418381a.

Thiel, Martin, Lars Gutow. "The Ecology of Rafting in the Marine Environment. II. The Rafting Organisms and Community." *Oceanography and Marine Biology: An Annual Review* 43 (July 2005): 279–418. doi:10.1201/9781420037449.ch7.

Thiel, Martin, Pilar A. Haye. "The Ecology of Rafting in the Marine Environment. III. Biogeographical and Evolutionary Consequences." *Oceanography and Marine Biology: An Annual Review* 44 (2006): 323–429. doi:10.1201/9781420006391.ch7.

What Is Cryptozoology? by Patricia Maldonado

Arment, Chad. *Cryptozoology: Science & Speculation*. 1st ed. Landisville, PA: Coachwhip Publications, 2004.

Bradford, Alina. "Facts About Ibex." *Live Science*. July 14, 2014. Accessed September 24, 2016. http://www.livescience.com/28102-ibex.html.

Chivian, Eric, and Aaron Bernstein. *Sustaining Life: How Human Health Depends on Biodiversity*. Oxford: Oxford University Press, 2008.

CNN. "More than 100,000 Rare Gorillas Found in Congo." *CNN*. August 6, 2008. Accessed May 17, 2016. http://www.cnn.com/2008/WORLD/africa/08/05/congo.gorillas/index.html?iref=topnews.

Coleman, Loren, and Patrick Huyghe. *The Field Guide to Lake Monsters, Sea Serpents and Other Mystery Denizens of the Deep*. New York, NY: Jeremy P. Tarcher/Putnam, 2003.

De Pastino, Blake. "Photo in the News: Giant Squid Captured, Filmed for First Time." *National Geographic*. December 22, 2006. Accessed September 20, 2016. http://news.nationalgeographic.com/news/2006/12/061222-giant-squid.html.

Gable, Andrew D. "Two Possible Cryptids From Precolumbian Mesoamerica." *Cryptozoology Review* 2, no. 1 (Summer 1997).

"Information/FAQ." *Australian Rare Fauna Research Association*. Accessed May 17, 2016. http://www.arfra.org/informationfaq.htm.

"Javan Tiger." *Global Tiger Initiative*. Accessed September 20, 2016. http://globaltigerinitiative.org/about-tigers/javan-tiger.

Lindsey, Susan Lyndaker, Mary Neel. Green, and Cynthia L. Bennett. *The Okapi: Mysterious Animal of Congo-Zaire*. Austin: University of Texas Press, 1999.

National Geographic Channel. "Jersey Devil Facts." *Nat Geo WILD*. September 24, 2013. Accessed September 20, 2016. http://channel.nationalgeographic.com/wild/the-monster-project/articles/jersey-devil-facts.

The Fantastic Annual Migration of North American Shorebirds by Anita K. Boyle

Boyer, Tim. *Audubon Society: Understanding Shorebirds: The Miracle of Migration*. Bellingham, WA: Whatcom Museum. March 22, 2016.

Ehrlich, Paul, David Dobki, and Darryl Wheye. "Shorebird Migration." *Birds of Standford*. 1988. http://web.stanford.edu/group/stanfordbirds/text/essays/Shorebird_Migration.html.

O'Brien, Michael, Richard Crossley, and Kevin Karlson. *The Shorebird Guide*. Boston: Houghton Mifflin Company, 2006.

Peterson, Roger T. *A Field Guide to Western Birds*, 2nd ed. Boston: Houton Mifflin Company, 1961.

Sibley, David A. *National Audobon Society: The Sibley Guide to Birds*. New York: Alfred A. Knopf, 2001.

The Day I Saw a Mountain Beaver by Natalie Dupille

Ostrom, Carol M. "The Pacific Northwest's Elusive Mountain Beaver." *Pacific NW Magazine*, February 8, 2009. http://www.seattletimes.com/pacific-nw-magazine/the-pacific-northwests-elusive-mountain-beaver.

Washington Department of Fish and Wildlife. "Living with Wildlife: Mountain Beavers." Accessed August 1, 2016. http://wdfw.wa.gov/living/mtn_beavers.html.

Babies Having Babies by Caitlin Hoffmeister and Lauren Norby

Agricultural Communications. "Potato Aphids." *North Carolina State University*. Last accessed September 15, 2016. http://ipm.ncsu.edu/AG295/html/potato_aphid.htm.

Miller, Renee. "Foxglove Aphids." *SFGate*. Last accessed October 9, 2016. http://homeguides.sfgate.com/citrus-aphid-life-cycle-72114.html.

"Rose Care: Aphids." *Ludwig's Roses*. Last accessed October 9, 2016. http://www.ludwigsroses.co.za/rose-care/aphids-2.

University of Florida. "Green Peach Aphid." *Featured Creatures*. Last accessed October 9, 2016. http://entnemdept.ufl.edu/creatures/veg/aphid/green_peach_aphid.htm.

West, Ron. "The Backyard Jungle XI: The Aphid Life Cycle." *Mother Earth News* (March/April 1985). http://www.motherearthnews.com/organic-gardening/aphid-life-cycle-zmaz85mazraw?PageId=2#ArticleContent.

Xiphactinus by Allison Bannister

Fish-Within-a-Fish. Second Floor, Permanent Collection, Fort Hays State University's Sternberg Museum of Natural History, Hays, Kansas.

Everhart, Mike. "*Xiphactinus audax*." *Oceans of Kansas*. Last modified August 1, 2013. http://oceansofkansas.com/xiphac.html.

Fort Hays University. "History." *Sternberg Museum of Natural History*. Last accessed October 22, 2016. http://sternberg.fhsu.edu/history.

Wikipedia. "Xiphactinus." *Wikipedia*. Last modified October 9, 2016. https://en.wikipedia.org/wiki/Xiphactinus.

Cordyceps by Simina Popescu

Castro, Joseph. "Zombie Fungus Enslaves Only Its Favorite Ant Brains." *Live Science*. September 9, 2014. http://www.livescience.com/47751-zombie-fungus-picky-about-ant-brains.html#sthash.a9CKZ4xu.dpuf.

Harmon, Katherine. "Fungus Makes Zombie Ants Do All the Work." *Scientific American*. July 31, 2009. http://www.scientificamerican.com/article/fungus-makes-zombie-ants.

———. "Undead-End: Fungus That Controls Zombie-Ants Has Own Fungal Stalker." *Scientific American*. November 8, 2012. http://www.scientificamerican.com/article/zombie-ant-fungus-parasite.

Hill, Kyle. "The Fungus that Reduced Humanity to *The Last of Us*." *Scientific American*. June 25, 2013. http://blogs.scientificamerican.com/but-not-simpler/the-fungus-that-reduced-humanity-to-the-last-of-us.

Loreto et al. "Fungus That Turns Hosts into Zombies Manipulates Ants to Die on Colony Doorstep." *IFL Science*. 2014. http://www.iflscience.com/plants-and-animals/fungus-turns-hosts-zombies-manipulates-ants-die-colony-doorstep.

The Degu by Bridget Comeau

Long, Chloe. *Degutopia*. Accessed October 5, 2016. http://www.degutopia.co.uk.

Welker, Wally, John Irwin Johnson, and Adrianne Noe. "Degu (Octodon degus) #66–40." *Neuroscience Library*. Accessed September 15, 2016. http://www.neurosciencelibrary.org/Specimens/rodentia/octdegu/index.html.

Extinction Superstar by Angela Boyle

Bolton, W. Draper. *Bolton's Mauritius Almanac, and Official Directory*. A. J. Tennant. 1852.

Cuppy, Will. "Dodos." *QI : Quite Interesting*. Accessed July 30, 2016. http://qi.com/infocloud/dodos.

Kasnoff, Craig. "Dodo Bird: An Extinct Species." *Bagheera*. 2016. Accessed June 18, 2016. http://www.bagheera.com/inthewild/ext_dodobird.htm.

Logic Media. *The Dodo Bird*. "History, Story and Resources for Dodo birds." Accessed June 18, 2016. http://www.dodobird.net.

Oxford University Museum of Natural History. "The Oxford Dodo." Accessed July 30, 2016. http://www.oum.ox.ac.uk/learning/pdfs/dodo.pdf.

Pappas, Stephanie. "Dodo Birds Weren't 'Dodos' After All." *Live Science*. February 24, 2016. Accessed June 18, 2016. http://www.livescience.com/53822-dodo-birds-were-smart.html.

Wikipedia. "Second Dutch Expedition to Indonesia." *Wikipedia*. Last modified October 2, 2015. https://en.wikipedia.org/w/index.php?title=Second_Dutch_Expedition_to_Indonesia&oldid=683720590.

———. "*Sideroxylon grandiflorum*." *Wikipedia*. Last modified September 21, 2015. https://en.wikipedia.org/w/index.php?title=Sideroxylon_grandiflorum&oldid=682123153.

Invasion of the Invasive Plants by Bonesteel

Kedzie-Webb, Susan, and Sheley, Roger L. Revised by Jane Mangold and Melissa Brown. "Houndstongue: Identification, Biology and Integrated Management." Self-Learning Resource from MSU Extension. Revised August 2009. http://store.msuextension.org/publications/AgandNaturalResources/MT199709AG.pdf.

USDA. "Invasive Plants." United States Department of Agriculture Forest Service. Accessed October 20, 2016. http://www.fs.fed.us/wildflowers/invasives.

———. *National Invasive Species Information Center*. Last modified September 28, 2016. https://www.invasivespeciesinfo.gov/index.shtml.

Why Great Blue Herons Eats Alone by Kami Koyamatsu

Cornell University. "Great Blue Heron." *The Cornell Lab of Ornithology*. 2015. https://www.allaboutbirds.org/guide/Great_Blue_Heron/id.

Kaufman, Kenn. "Great Blue Heron." *National Audubon Society*. 2016. http://www.audubon.org/field-guide/bird/great-blue-heron.

Sibley, David Allen. *The Sibley Guide to Birds, Second Edition*. New York, NY: Random House, LLC, 2014.

The Sounds and the Silence of the 17-Year Cicada
by Sophie Argetsinger

Cicada Mania. Accessed July 14, 2016. http://www.cicadamania.com.

Cooley, John R. *Periodical Cicadas*. October 6, 2016. Accessed July 14, 2016. http://www.magicicada.org.

Roach, John. "Cicada Swarm Proves a Feast for Predators." *National Geographic*. May 24, 2004. Accessed July 14, 2016. http://news.nationalgeographic.com/news/2004/05/0524_040524_cicadaterminator.html.

Roza, Greg. *The Bizarre Life Cycle of a Cicada*: 4, 8–14. New York: Gareth Stevens Publishing, 2013.

Sexton, Colleen. *World of Insects: Cicadas*: 10–14. Minneapolis: Bellwether Media, 2007.

The Ears Have It by Payden McRoberts

Gupta, Sanjay. "If We Are What We Eat, Americans Are Corn and Soy." CNN. September 22, 2007. http://www.cnn.com/2007/HEALTH/diet.fitness/09/22/kd.gupta.column.

Iltis, Hugh H. "From Teosinte to Maize: The Catastrophic Sexual Mutation" *Science* 222, no. 4626 (November 25, 1983). doi:10.1126/science.222.4626.886.

Philpott, Tom. "A Reflection on the Lasting Legacy of 1970s USDA Secretary Earl Butz." *Grist*. February 8, 2008. http://grist.org/article/the-butz-stops-here.

Pollan, Michael. *The Omnivore's Dilemma: A Natural History of Four Meals*. London: Penguin, 2007.

Warman, Arturo. *Corn and Capitalism: How a Botanical Bastard Grew to Global Dominance*. The University of North Carolina Press, 2003.

Monotremes by Abe Olson and Angela Boyle

Wikipedia. "Monotreme." *Wikipedia*. Last modified October 22, 2016. https://en.wikipedia.org/wiki/Monotreme.

———. "Prototheria." *Wikipedia*. Last modified October 14, 2016. https://en.wikipedia.org/wiki/Prototheria.

———. "Metatheria." *Wikipedia*. Last modified September 24, 2016. https://en.wikipedia.org/wiki/Metatheria.

———. "Eutheria." *Wikipedia*. Last modified September 26, 2016. https://en.wikipedia.org/wiki/Eutheria.

———. "Platypus." *Wikipedia*. Last modified October 22, 2016. https://en.wikipedia.org/wiki/Platypus.

———. "Echidna." *Wikipedia*. Last modified October 20, 2016. https://en.wikipedia.org/wiki/Echidna.

Coulee Crowns by Laura Marie Madden

Tames, Stacey. "Western/Plains Cottonwood (*Populus deltoides* ssp. *monilifera*)." Alberta Agriculture and Forestry. March 30, 2016. http://www.agric.gov.ab.ca/app21/forestrypage?cat1=Trees&cat2=Western%2FPlains%20Cottonwood.

Bain, John, June Flanagan, and Job Kuijt. "Common Coulee Plants of Southern Alberta." 2014. http://www.npss.sk.ca/docs/2_pdf/Common_Coulee_Plants_of_Southern_Alberta_-_31_03_2014.pdf.

Ball, John, Dave Erikson, and Christopher Lewandowski. "Prairie Forests." *The Nature Source*, Northern State University. Last Accessed September 15, 2016. http://www3.northern.edu/natsource/HABITATS/Prairi1.htm.

Egan, Brian, Carmen Cadrin, and Syd Cannings. "Cottonwood Riparian Ecosystems of the Southern Interior." British Columbia Minstry of Environment, Lands and Parks. August 1997. http://www2.gov.bc.ca/assets/gov/environment/plants-animals-and-ecosystems/species-ecosystems-at-risk/brochures/cottonwood_riparian_ecosystems_southern_interior.pdf.

Helen Schuler Nature Center. "Coulees and Cottonwoods." *Nature Field Guide for Lethbridge*. Last accessed September 15, 2016. http://www.lethbridge.ca/Things-To-Do/Nature-Centre/Documents/HSNC%20Field%20Guide%20singlepages.pdf.

"Cottonwood Forests." *Multisar*. October 2015. http://multisar.ca/wp-content/uploads/2015/10/cottonwoods2.pdf.

Police Point Park Interpretive Center. Medicine Hat Interpretive Program. Last accessed September 15, 2016. http://www.natureline.info.

Rivers West. "Forgotten Forests of the Prairies." *Routes on the Red*. 2004. http://www.routesonthered.ca/media/documents/brochure-forests.pdf.

University of Lethbridge. "The Cottonwood Biome." Virtual Field Tours. Last accessed September 15, 2016. http://www.uleth.ca/vft/Oldman_River/Cottonwood.html.

Weiler, Jason. "Conserving the Grasslands of Southern Alberta." Canadian Parks and Wilderness Society, Southern Alberta Chapter. September 2010. http://cpaws-southernalberta.org/upload/Conserving_the_Grasslands_of_Southern_Alberta_Report_FINAL_June11.pdf.

The Grasshopper Mouse by Jerel Dye

Encyclopaedia Britannica. "Grasshopper Mouse." *Encyclopaedia Britannica*. Accessed September 15, 2016. https://www.britannica.com/animal/grasshopper-mouse.

Reardon, Sara. "Mouse Eats Scorpions and Howls at the Moon." *Zoologger* (blog), *New Scientist*. January 11, 2013. https://www.newscientist.com/article/dn23072-zoologger-mouse-eats-scorpions-and-howls-at-the-moon.

Rose, Meghan, and Science News. "The Tiny Southern Grasshopper Mouse Can Eat Scorpions without Feeling Any Pain." *Washington Post*, Health and Science. November 4, 213. https://www.washingtonpost.com/national/health-science/the-tiny-southern-grasshopper-mouse-can-eat-scorpions-without-feeling-any-pain/2013/11/04/a84bbf9a-425d-11e3-a751-f032898f2dbc_story.html.

Texas Tech University. "Mearns' Grasshopper Mouse." *The Mammals of Texas—Online Edition*. Accessed September 15, 2016. http://www.nsrl.ttu.edu/tmot1/onycaren.htm.

——. "Northern Grasshopper Mouse." *The Mammals of Texas—Online Edition*. Accessed September 15, 2016. http://www.nsrl.ttu.edu/tmot1/onycleuc.htm.

Yong, Ed. "This Mouse Turns Agonising Scorpion Venom into a Painkiller." *Not Exactly Rocket Science* (blog). October 24, 2013. National Geographic, Phenomena: A Science Salon. http://phenomena.nationalgeographic.com/2013/10/24/this-mouse-turns-agonising-scorpion-venom-into-a-painkiller.

Zhang, Sarah. "Rodent Immune to Scorpion Venom." *Nature*, News. October 24, 2013. http://www.nature.com/news/rodent-immune-to-scorpion-venom-1.14014.

Living Large by Hedj

Castro, José I. "Orectolobiformes (Carpet Sharks)." *Grzimek's Animal Life Encyclopedia*. Ed. Michael Hutchins, et al. 2nd ed. Vol. 4: Fishes I. Detroit: Gale, 2004. 105–112. Gale Virtual Reference Library. June 29, 2016.

Lister, Adrian. "Proboscidea (Elephants)." *Grzimek's Animal Life Encyclopedia*. Ed. Michael Hutchins, et al. 2nd ed. Vol. 15: Mammals IV. Detroit: Gale, 2004. 161–175. Gale Virtual Reference Library. June 29, 2016.

"Blue Whales, *Balaenoptera musculus*." *MarineBio Conservation Society*. Accessed June 10, 2016. http://marinebio.org/species.asp?id=41.

Mazzetta, Gerardo V., Per Christiansen, and Richard A. Fariña. "Giants and Bizarres: Body Size of Some Southern South American Cretaceous Dinosaurs." *Historical Biology* 16, no. 2–4 (January 2004): 71–83.

Mitton, Jeffry Bond and Michael C. Grant. "Genetic Variation and the Natural History of Quaking Aspen." *BioScience* 46. no. 1 (January 1996): 25–31. doi:10.2307/1312652.

Rogers, Paul C. "Saving Pando." *Natural History* 124 (2016): 32–37.

Volk, Thomas J. *The Humongous Fungus—Ten Years Later*. Accessed June 12, 2016. http://botit.botany.wisc.edu/toms_fungi/apr2002.html.

Monsters Unmasked by Patricia Maldonado

Moriceau, Jean-Marc. "Man and Wolf: 200 Years of History." *Histoires Des Loups*. March 2014. Accessed March 13, 2016. http://www.unicaen.fr/homme_et_loup/_en/index.php.

Nickell, Joe. "Mothman Revisited: Investigating on Site." *Skeptical Briefs* 12, no. 4 (December 2002).

Olson, Eric R. "The World's Most Endangered Frogs." PBS. June 17, 2014. Accessed September 24, 2016. http://www.pbs.org/wnet/nature/fabulous-frogs-the-worlds-most-endangered-frogs/8917.

Radford, Benjamin. *Tracking the Chupacabra: The Vampire Beast in Fact, Fiction, and Folklore*. Albuquerque: University of New Mexico Press, 2011.

Thompson, Richard H. *Wolf-hunting in France in the Reign of Louis XV: The Beast of the Gévaudan*. Lewiston, NY: E. Mellen Press, 1991.

Mating Support Group by Aurora Melchor

Arnqvist, Göran, and Locke Rowe. *Sexual conflict*. Princeton, NJ: Princeton University Press, 2005.

Michiels, N. K. and Newman, L. J. "Sex and Conflict in Hermaphrodites." *Nature* 391, no 647 (February 12, 1998). doi:10.1101/cshperspect.a017673.

Pfennig, David W. "Ecology and Evolution Among Diverse Taxa." Review of Cannibalism by M. A. Elgar and B. J. Crespi (eds.), Oxford University Press: 1992. *Journal of Evolutionary Biology* 7, no. 1 (January 1994). doi:10.1046/j.1420-9101.1994.7010121.x.

Pogány, Ákos, István Szentirmai, Jan Komdeur, and Tamás Székely. "Sexual Conflict and Consistency of Offspring Desertion in Eurasian Penduline Tit *Remiz pendulinus*." *BMC Evolutionary Biology* 8, no. 1 (September 2008): 242. doi:10.1186/1471-2148-8-242.

Sinervo, B., and C. M. Lively. "The Rock–Paper–Scissors Game and the Evolution of Alternative Male Strategies." *Nature* 380, no. 6571 (March 21, 1996): 240–243. doi:10.1038/380240a0.

Filial Cannibalism by Vicky Yuh

Mehlis, Marion, Theo C. M. Bakker, Leif Engqvist, and Joachim G. Frommen. "To Eat or Not to Eat: Egg-Based Assessment of Paternity Triggers Fine-Tuned Decisions about Filial Cannibalism." *Proceedings: Biological Sciences* 277, no. 1694 (April 21, 2010): 2627–35. doi:10.1098/rspb.2010.0234.

Vallon, Martin, Christina Grom, Nadine Kalb, Dennis Sprenger, Nils Anthes, Kai Lindström, and Katja U. Heubel. "You Eat What You Are: Personality-Dependent Filial Cannibalism in a Fish with Paternal Care. *Ecology and Evolution* 6, no. 5 (March 2016): 1340–52. doi:10.1002/ece3.1966.

Moose Mortality by Stephanie Zuppo

Allardyce, Mark. "Moose Brain Worm Moose Sickness or Parelaphostrongylosis." *All About Moose*. Last accessed September 15, 2015. http://www.all-about-moose.com/moose-brain-worm.html.

Benko, Jessica. "What's Killing Minnesota's Moose?" *On Earth*. July 15, 2013. http://archive.onearth.org/articles/2013/07/climate-change-could-wipe-out-minnesotas-iconic-moose.

Carey, John. "The Deepening Mystery of Moose Decline." National Wildlife Federation. September 29, 2014. http://www.nwf.org/news-and-magazines/national-wildlife/animals/archives/2014/moose.aspx.

Dell'Amore, Christine. "What's a Ghost Moose? How Ticks Are Killing an Iconic Animal." *National Geographic*. June 1, 2015. http://news.nationalgeographic.com/2015/06/150601-ghost-moose-animals-science-new-england-environment.

Mathewson, Samantha. "Minnesota Moose: Researchers Reveal What's Causing Population Declines." *Nature World News*. January 27, 2016. http://www.natureworldnews.com/articles/19558/20160127/minnesota-moose-researchers-reveal-whats-causing-population-declines.htm.

Morse, Susan C. "Declining Moose Populations: What Does the Future Hold?" *Northern Woodlands* Spring 2015 (April 10, 2015). http://northernwoodlands.org/articles/article/moose-decline.

Natural Resource Institute. "Moose in Minnesota." *University of Minnesota*. 2004. http://nrri.d.umn.edu/moose/information/diseases.html.

"A North Woods Icon at Risk." *National Wildlife Federation*. Accessed October 2016. https://www.nwf.org/Wildlife/Threats-to-Wildlife/Global-Warming/Effects-on-Wildlife-and-Habitat/Moose.aspx.

Rines, Kristine. "New Hampshire's Moose Population vs. Climate Change." *University of New Hampshire*. Last accessed September 15, 2016. https://extension.unh.edu/resources/files/Resource003858_Rep5484.pdf.

Robbins, Jim. "Moose Die-Off Alarms Scientists." *New York Times*. October 14, 2013. http://www.nytimes.com/2013/10/15/science/earth/something-is-killing-off-the-moose.html.

Tracy, Paula. "Most Moose Calves in Study Died This Winter." *WMUR*. June 7, 2015. http://www.wmur.com/escape-outside/most-moose-calves-in-study-died-this-winter/33352380.

Wildlife Management Institute. "States Initiating Research on Moose Declines; Minnesota Halts Hunt." *Outdoor News Bulletin* 67, no. 2 (February 15, 2013). https://wildlifemanagement.institute/node/6057.

The Immortal Jellyfish by Kelly Chen

AMNH. "The Immortal Jellyfish." *American Museum of Natural History*. May 4, 2015. http://www.amnh.org/explore/news-blogs/on-exhibit-posts/the-immortal-jellyfish.

Barras, Colin. "The Animals and Plants That Can Live Forever." *BBC Earth*. Last modified June 19, 2015.http://www.bbc.com/earth/story/20150622-can-anything-live-forever.

Boero, Ferdinando. "Everlasting Life: The 'Immortal' Jellyfish." *The Biologist* 63, no. 3 (2016): 16–19. https://thebiologist.rsb.org.uk/biologist/158-biologist/features/1510-everlasting-life-the-immortal-jellyfish.

Deza, Michel Marie and Elena Deza. *Encyclopedia of Distances*. Heidelberg: SpringerVerlag, 2009, 2013.

Ericfein. "Immortal Animals and the Human Link ." *Wondergressive*. Last modified July 7, 2013. http://wondergressive.com/hydra-immortality-linked-to-human-lifespan.

Ma, Hongbao and Yan Yang. "Turritopsis nutricula." *Nature and Science* 8, no. 2 (2010). http://www.sciencepub.net/nature/ns0802/03_1279_hongbao_turritopsis_ns0802_15_20.pdf.

Rich, Nathaniel. "Can a Jellyfish Unlock the Secret of Immortality?" *New York Times Magazine*. November 28, 2012. http://www.nytimes.com/2012/12/02/magazine/can-a-jellyfish-unlock-the-secret-of-immortality.html.

Than, Ker. "'Immortal' Jellyfish Swarm World's Oceans." *National Geographic News*. January 29, 2009. http://news.nationalgeographic.com/news/2009/01/090130-immortal-jellyfish-swarm.html.

Flightless Rails and Island Conservation by G Pike

Cruce, Jennifer. "Fencing in Forest on Guam." National Wildlife Refuge System. Last modified July 1, 2013. https://www.fws.gov/refuges/refugeupdate/JulAug_2013/guam-fencing.html.

Kirchman, Jeremy J. "Speciation of Flightless Rails on Islands: A DNA-Based Phylogeny of the Typical Rails of the Pacific." *The Auk* 129, no. 1 (January 2012): 56–69. http://www.bioone.org/doi/abs/10.1525/auk.2011.11096.

Government of New Zealand. "Takah?" *New Zealand Department of Conservation*. Accessed July, 2016. http://www.doc.govt.nz/takahe.

——. "Weka: Native land birds: Conservation." *New Zealand Department of Conservation*. Accessed July, 2016. http://www.doc.govt.nz/nature/native-animals/birds/birds-a-z/weka.

Platt, John R. "Memorializing the Wake Island Rail: An Extinction Caused by War." *Scientific American*. May 25, 2015. http://blogs.scientificamerican.com/extinction-countdown/memorial-day-extinction.

Taylor, B. "Guam Rail (*Hypotaenidia owstoni*)." *Handbook of the Birds of the World Alive*. Lynx Edicions, Barcelona. 2016. http://www.hbw.com/node/53616.

Wikipedia. "Rallidae." *Wikipedia*. Accessed July, 2016. https://en.wikipedia.org/wiki/Rallidae.

———. "Wake Island rail." *Wikipedia*. Last modified May 31, 2015. Accessed July, 2016. https://en.wikipedia.org/wiki/Wake_Island_rail.

How I Met Your Father by Iris Yan

Albo, Maria J., and Fernando G. Costa. "Nuptial Gift-Giving Behaviour and Male Mating Effort in the Neotropical Spider *Paratrechalea ornata* (Trechaleidae)." *Animal Behavior* 79, no. 5 (May 2010): 1031–36. doi:10.1016/j.anbehav.2010.01.018.

Hansen, Line Spinner, Sofía Fernández González, Søren Toft, and Trine Bildea. "Thanatosis as an Adaptive Male Mating Strategy in the Nuptial Gift–Giving Spider *Pisaura mirabilis*." *Behavioral Ecology* 19, no. 3 (February 17, 2008): 546–51. doi:10.1093/beheco/arm165.

Stålhandske, Pia. "Nuptial Gifts of Male Spiders Function as Sensory Traps." *Proceedings of the Royal Society B: Biological Sciences* 269, no. 1494 (May 7, 2002): 905–8. doi:10.1098/rspb.2001.1917.

Welsh, Jennifer. "Male Spiders Let Mates Eat Them for Kids' Sake." *Live Science*. December 23, 2011. http://www.livescience.com/17616-spider-sexual-cannibalism-offspring.html.

Meet Percy! by EJ Landsman

ASPCA. "Position Statement on Declawing Cats." *American Society for the Prevention of Cruelty to Animals*. Accessed September 2016. http://www.aspca.org/about-us/aspca-policy-and-position-statements/position-statement-declawing-cats.

HSUS. "Cat Chat: Understanding Feline Language." *The Humane Society of the United States*. Accessed September 2016. http://www.humanesociety.org/animals/cats/tips/cat_communication.html.

Fehér, György and András Szunyoghy. *Cyclopedia Anatomicae: More Than 1,500 Illustrations of the Human and Animal Figure for the Artist*. New York: Black Dog & Leventhal Publishers, Inc., 1996.

Cute or Deadly: Starlings by Abe Olson and Angela Boyle

Wikipedia. "Common Starling." *Wikipedia*. Last modified October 23, 2016. https://en.wikipedia.org/wiki/Common_starling.

Barras, Scott C., Sandra E. Wright, and Thomas E. Seamans. "Blackbird and Starling Strikes to Civil Aircraft in the United States, 1990–2001." University of Nebraska. September 2003. http://digitalcommons.unl.edu/cgi/viewcontent.cgi?article=1194&context=icwdm_usdanwrc.

"Worst Bird Strike, 1960." *Celebrate Boston*. Last accessed October 23, 2016. http://www.celebrateboston.com/disasters/logan-electra-bird-strike-1960.htm.

Carlic, Steve. "Introducing America's Most Hated Bird: The Starling." *The Post-Standard*, September 7, 2009. http://www.syracuse.com/today/index.ssf/2009/09/introducing_americas_most_hate.html.

Fox by Wm Brian MacLean

Aggas, Edward, trans. "The Contre-League [Microform]: And Ansvvere to Certaine Letters Sent to the Maisters of Renes, by One of the League Who Termeth Himselfe Lord of the Valley of Mayne, and Gentleman of the Late Duke of Guizes Traine. Faithfully Translated into English by E.A." London, England: Printed by Iohn Wolfe, 1589. https://searchworks.stanford.edu/view/9458085.

Arnold, Caroline and Richard Hewett. *Fox*, 25–28, 44–46. New York, NY: Morrow Junior Books, 1996.

Bruce, Elyse. "Fox in the Henhouse," *Historically Speaking*. March 6, 2013. Accessed June, 2016. https://idiomation.wordpress.com/2013/03/06/fox-in-the-henhouse.

Forsyth, Adrian. "Foxes & Wolves." In *Mammals of North America: Temperate and Arctic Regions,* 166–181. Buffalo, NY: Firefly Books, 1999.

Government of Canada. "Gray Fox." SARA (Species at Risk Assessment), Species at Risk Public Registry. Last modified June 22, 2016. http://www.registrelep-sararegistry.gc.ca/species/speciesDetails_e.cfm?sid=157.

———. "Swift Fox." SARA (Species at Risk Assessment), Species at Risk Public Registry. Last modified June 22, 2016. http://www.registrelep-sararegistry.gc.ca/species/speciesDetails_e.cfm?sid=140.

Government of Ontario. "Fish and Wildlife Conservation Act, 1997, S.O. 1997, c. 41." https://www.ontario.ca/laws/statute/97f41.

Leach, Michael. "Threats." In *Fox (Animal Neighbours),* 24–27. New York, NY: Rosen Publishing Group, 2009.

Macdonald, David W., ed. "Foxes." In *The Princeton Encyclopedia of Mammals,* 604–611. Princeton, NJ: Princeton University Press, 2009.

Rinik, Jan, and Martin Rinik. "Autumn—Nature's Palette." In *Algonquin Park—A Portrait,* 170–171. Halifax, NS: Formac Publishing Company, 2014.

A Pictorial Anatomy of the Denizens of Manhattan
by Kriota Willberg

Carothers, J., and Cabrillo College. "Bird Anatomy." *EcoWebSite.* Accessed July 6, 2016. http://www.cabrillo.edu/~jcarothers/ecology/EcoWebSite/lab/Bird-Anatomy.pdf.

F., Chris. "Panting in Racing Pigeons—When Is It Normal and When Is It Not, Part 1." *The Pigeon Insider.* Accessed June 26, 2016.http://www.pigeonracingpigeon.com/whats-new/panting-in-racing-pigeons—when-is-it-normal-and-when-is-it-not-part-1.

Fowler, Denver W., Elizabeth A. Freedman, and John B. Scannella. "Predatory Functional Morphology in Raptors: Interdigital Variation in Talon Size Is Related to Prey Restraint and Immobilisation Technique." *PLosONE.* November 25, 2009. Accessed July 6, 2016. doi:10.1371/journal.pone.0007999.

Gotham Coyote Project. "Urban Conservation: The Northeastern Coyote as a Flagship Species." *Gotham Coyote.* April 30, 2016. http://www.gothamcoyote.com/news/urban-conservation-the-northeastern-coyote-as-a-flagship-species.

———. "What Do You Eat?" *Gotham Coyote.* Accessed July 5, 2016. http://www.gothamcoyote.com/what-do-coyotes-eat.html.

Richardson, Lance. "New York Needs Coyotes." *Slate.* July 31, 2015. Accessed July 10, 2016. http://www.slate.com/articles/health_and_science/science/2015/07/coyotes_in_new_york_and_chicago_urban_ecology_of_rats_geese_deer_feral_cats.html.

Ritchison, Gary. *BIO 554/754 Ornithology, Avian Respiration.* Eastern Kentucky University. NA. Accessed June 26, 2016. http://people.eku.edu/ritchisong/birdrespiration.html.

Smith, David G., and Michael P. Schenk. *A Dissection Guide & Atlas to the Rat.* Englwood, Colorado: Morton Publishing Company, 2001.

Stanley, Susan M. "A Study of Heat Stroke and Heat Exhaustion in the Dog." *Iowa State University Veterinarian* 42, no. 1 (1980): Article 8.

Wikipedia. "Ungulate." *Wikipedia.* Last update May 29, 2016. https://en.wikipedia.org/wiki/Digitigrade.

Preserving the Silurian Age by Kes Bradley and David Kirkham

Black Country Bugle. "Heraldic Dudley Postcard with News from Edwardian Days. *Black Country Bugle.* April 4, 2014. http://www.blackcountrybugle.co.uk/heraldic-postcard-news-edwardian-days/story-20898145-detail/story.html.

Dudley Museum and Art Gallery. St James's Rd, Dudley), United Kingdom.

Black Country Geological Society. *Geology Matters.* Last accessed October 5, 2016. http://geologymatters.org.uk.

Hamblin, R.J.O. *Geological Handbook for the Wren's Nest National Nature Reserve.* Geology and Physiography Section, Nature Conservancy Council, 1978.

Nature Conservancy Council. *Wren's Nest National Nature Reserve.* 1967.

Wren's Nest Nature Reserve. Wrens Hill Rd, Dudley, United Kingdom.

A Tale of the Horsetail by Anita K. Boyle

Armstrong, Joseph E. *How the Earth Turned Green: A Brief 3.8-Billion-Year History of Plants.* Chicago: University Of Chicago Press, 2015.

Cullina, W. *Native Ferns, Moss and Grasses.* Boston: Houghton Mifflin Company, 2008.

Germplasm Resource Information Network. *Global Plant Genebank Information Management System.* 2005. https://npgsweb.ars-grin.gov.

Pojar, Jim, and Andy MacKinnon. *Revised Plants of the Pacific Northwest Coast, Washington, Oregon British Columbia and Alaska.* Vancouver: Lone Pine, 1994.

Pratt, Anne. *The Ferns of Great Britain and Their Allies, the Club-Mosses, Pepperworts, and Horsetails,* 2nd ed. London: printed for the Society for Promoting Christian Knowledge, 1871.

Jackie. "Wild Horse Tail." *Recipes from the Wild: All Kinds of Recipes.* 2013. https://recipesfromthewild.wordpress.com/wild-horse-tail.

Iberian Lynx by Bernardo Dias

Rodríguez, A., and Calzada, J. "Lynx pardinus." *The IUCN Red List of Threatened Species.* 2015. doi:10.2305/IUCN.UK.2015-2.RLTS.T12520A50655794.

Vargas, A., I. Sánchez, F. Martínez, A. Rivas, J. A. Godoy, E. Roldán, M. A. Simón, et al. "The Iberian Lynx *Lynx pardinus* Conservation Breeding Program." *International Zoo Yearbook* 42. no. 1 (April 2008): 190–198. doi:10.1111/j.1748-1090.2007.00036.x.

American Museum of Natural History. "Sperm Whales' Amazing Adaptations." Accessed April 10, 2016. http://www.amnh.org/explore/news-blogs/on-exhibit-posts/sperm-whales-amazing-adaptations.

INDEX

Biographies

Megan Archer is a cartoonist and software engineer living in Seattle, Washington. She's currently trying to grow a jungle on her balcony.
MeganArcherComics.com
MeganArcherComics@gmail.com

Sophie Argetsinger is an illustrator, book designer, hiker, and crocodile-level Reptile Scout. She lives in Holyoke, Massachusetts with her wife, Jozen, and her daughter, Addie.
sophieargetsinger.com
sargetsinger@gmail.com

Allison Bannister is a cartoonist originally from Kansas. She loves cats, tea, and dinosaurs and is currently pursuing a PhD in communication.
basictelepathy.com
allisoncbannister@gmail.com

Moss Bastille spends his time making comics and trying to trick people into thinking he's an adult. He was raised by squirrels.
moth-ire.tumblr.com
mbastille@coa.edu

Bonesteel. Cartoonist. Father. Horror enthusiast. MFA in cartooning from The Center for Cartoon Studies. Amateur chef. Baller. Northern boy at heart.
killerinkcomics.com
killerink_13@hotmail.com

Angela Boyle is a natural science illustrator, cartoonist, and editor. Her favorite animal is the Brazilian tapir. She owns two corgis, Nisa and Ernie.
AngelaBoyle.FlyingDodoStudio.com
angela@flyingdodopublications.com

The environment around Anita K. Boyle winds up in her art, whether as paper, an assemblage, or in a poem.
EgressStudio.com
akboyle@egressstudio.com

After finishing a degree in screenwriting, Kes Bradley made the natural career progression of becoming a support worker. Keen to get back into writing, he is currently studying for a master's in his spare time and spends the days scribbling scripts whilst riding the bus to and from work.

Kelly Chen is a student studying illustration at Emily Carr University who looks up to a sponge. She hopes to master the ability to "soak up" knowledge.
Scarlet-Wings-Kaili.deviantart.com
scarletwingskaili@gmail.com

Bridget Comeau. Gnome queen, mushroom enthusiast, lover of all woodland creatures. CCS 2015.
bridgetcomeau.tumblr.com

Pup wolf wrangler, Jackal rescuer, and lynx watcher. Bernardo O. Dias likes to read and draw; sometimes people pay him for the last one.
www.facebook.com/bernardoodias
bernardo.nevoa@oliveiradias.eu

Natalie Dupille is a cartoonist, illustrator, and educator based in the Pacific Northwest with a love for travel, nature, and working with children.
nataliedupille.com
nataliedeee@gmail.com

Jerel Dye is a cartoonist and illustrator in the Boston area. His first graphic novel, *Pigs Might Fly*, is due to be released in 2017, by First Second.
jereldye.com

Kit Fraser is an illustrator and comic artist based in New York. When she's not doodling, she's out in the woods with her wild dog.
kitfraser.com
kitkfraser@gmail.com

Kelly Fernandez was born and raised in Queens, New York. She loves making comics and screen prints. Her hobbies are swimming and looking for lizards.
kelly-fernandez.com
kelly@kellyfernandezart.com

Catherine Garbarino is a Kansas artist living in Vermont. She loves the mystical and macabre.
catherinegarbarino.tumblr.com.

Hedj is a cartoonist from rural New York. When not drawing, Hedj enjoys cooking, costuming, catching salamanders, and crying about cartoons.
vesparts.tumblr.com
primitiverobot@gmail.com

Caitlin Hoffmeister is a filmmaker and artist from Missoula, Montana. She's the producer of the YouTube series *SciShow*. This is her first comic.

David Kirkham has contributed cartoons to *Run of the Mill*, *Villager Magazine*, and the *Forbidden Planet* website. Most recently he has been developing a project with the BBC. He is the author of the semi-autobiographical comic *Tripped Electric* at trippedelectric.tumblr.com.
www.facebook.com/DaveyDoodlebug
trippedelectric@gmail.com

Kevin Kite and Michelle McCauley create the science and culture comic *Hurry Up Please It's Time*. They live in the green hills of Vermont with two kids, two cats, some skunks, and several thousand bees.
hurryuppleaseitstime.com
klkite@gmavt.net

Ethan Kocak lives in upstate New York with his wife and son. He has eight aquariums and a webcomic about a crime-fighting axolotl.
blackmudpuppy.com
ethan@blackmudpuppy.com

Kami Koyamatsu is a natural science illustrator and hopes to use her art to motivate conservation efforts by piquing people's interests in nature.
kamikoyamatsu.com
kamikoy@uw.edu

Kate Lacour is a New-Orleans-based cartoonist and amateur entomologist. She is the caretaker of one bearded dragon, three tortoises, and two humans.
www.katelacour.com

EJ Landsman is an artist and educator working and living in Seattle, Washington. They make comics about nature, gender, living with mental illness, and other normal things.
ejlandsman.com
ejlandsman at gmail.com

Lisa Lowndes is a writer of horror, romance, science, and occupational safety documents. She would play Dungeons and Dragons constantly if she could.
alchemist-and-druid.tumblr.com
alchemist.and.druid@gmail.com

Rural-Ontario cartoonist Wm Brian MacLean tries to inspire non-plastic, non-CGI interests in his grandson. Variously, Brian loves beer, grump, and <4.
roostertree.ca/biblio
roostertree@gmail.com

Laura Marie Madden is a cartoonist and illustrator from Alberta, Canada. She loves monsters, myths, and hockey, and hopes to make comics about all of it.
lauramariemadden.com
lauramariemadden@gmail.com

Her friends begged her to stop talking about Mothman, so Patricia "Sunny" Maldonado decided to inflict her love of cryptids and monsters on you poor unsuspecting readers.
bully-goat.tumblr.com
hurlyberly@gmail.com

Melissa McGee, when not creating art in various media or engrossed deeply in a book, pets dogs professionally. She lives near Toronto.
alchemist-and-druid.tumblr.com
alchemist.and.druid@gmail.com

Payden McRoberts has always admired the power of stories. He is a jack of all trades, and master of none. He loves making comics, eating poutine, and playing with his 2-year-old son.
twitter.com/paydirt21
payden.mcroberts@gmail.com

Aurora Melchor is a comicker and illustrator based in London. Comics are her favorite method of expression, and she's exploring its different applications.

Shashwat Mishra is an Indian artist who has seen the royal Bengal tiger in the jungle on two different occasions and lived to tell the tale. He also loves a good steak.
Shashmishra.tumblr.com
shashcomics@gmail.com

Lauren Norby is an artist and storyteller making comics, videos and music. He lives in Missoula, Montana, USA.
l-t-n.tumblr.com
laurennorby@gmail.com

TOM O'BRIEN is a 2014 graduate of the Center for Cartoon Studies. His previous works include *Rita*, *Madam Geneva*, and *Pouty Pope*. He lives in Troy, New York.
tomobriencomics.com
tomobriencomics@gmail.com

G PIKE is an author/illustrator from Indiana who has been making comics since 2000. Also an avid birdwatcher and world-traveler.
artofgpike.com
gpikestudios@gmail.com

SIMINA POPESCU is a 19-year-old Romanian illustrator, currently based in London, studying graphic communication design.
siminapopescuart.com
simina.popescu356@gmail.com

KELLY M. RICKER is of a nerdy persuasion with a deep love for Star Wars and all things animated. She is an artist for mobile games and often babysits her sister's three dogs.
artofkmricker.tumblr.com
kmricker@gmail.com

WILLIAM SCAVONE, a medical illustrator, lives in Vermont, with his wife and kids, cat, dog, goats, ducks, chickens, and bees. He's taught life drawing at The Center for Cartoon Studies since 2008.
www.kestrelstudio.com
kestrelstudio@gmail.com

SPRATTY, or Sarah Spratlin, lives in Drexel Hill, Pennsylvania, and spends her days drawing pictures, making t-shirts, cuddling snakes, and studying martial arts.
spratty-duck.tumblr.com

ROSS WOOD STUDLAR is a park ranger at Yellowstone and a graduate of The Center for Cartoon Studies. He practices martial arts and cooks a wholesome pasta salad.
rosswoodstudlar.blogspot.com
rwoodstudlar@gmail.com

KIM SUCHER is an illustrator currently residing in Buffalo, New York. Her passions are video games, comics, and nature—especially birds!
ksucher.weebly.com
DestryHawk@gmail.com

ALYSSA SUZUMURA is an invertebrate taxonomist specializing in insect identification at the University of Washington. She spends a lot of time looking at tiny things in a microscope and occasionally goes outside and gets muddy while looking for critters.
alsuzumura.tumblr.com
asuzu5@gmail.com

Through graphic narratives, teaching, performance, and needlework, KRIOTA WILLBERG focuses on the intersection between body sciences and creative practice.
KriotaWelt.blogspot.com
kriota@earthlink.com

IRIS YAN is a Brazilian-Chinese cartoonist who is a professionally trained aura reader. She believes life is humorous and likes to make funny comics.
pigsinmaputo.blogspot.com
pigsinmaputo@gmail.com

VICKY YUH is an illustrator who is inspired by worlds of fantasy and magical realism. She is currently working on *Luna*, her first long-form webcomic.
www.vickyuh.com
vickviggles@gmail.com

STEPHANIE ZUPPO is a cartoonist and educator living in Burlington, Vermont. She writes and draws the series *Belchville, VT*, which contains a lot of moose.
www.stephaniezuppo.com
stephanie.zuppo@gmail.com

www.ingramcontent.com/pod-product-compliance
Lightning Source LLC
LaVergne TN
LVHW060021210926
848481LV00013B/3761
9780997011128